Studies in Applied Philosophy, Epistemology and Rational Ethics

Volume 13

For further volumes:
http://www.springer.com/series/10087

About this Series

Studies in Applied Philosophy, Epistemology and Rational Ethics (SAPERE) publishes new developments and advances in all the fields of philosophy, epistemology, and ethics, bringing them together with a cluster of scientific disciplines and technological outcomes: from computer science to life sciences, from economics, law, and education to engineering, logic, and mathematics, from medicine to physics, human sciences, and politics. It aims at covering all the challenging philosophical and ethical themes of contemporary society, making them appropriately applicable to contemporary theoretical, methodological, and practical problems, impasses, controversies, and conflicts. The series includes monographs, lecture notes, selected contributions from specialized conferences and workshops as well as selected PhD theses.

Advisory Board

Matthew Cotton

Ethics and Technology Assessment: A Participatory Approach

 Springer

Matthew Cotton
Faculty of Social Sciences
University of Sheffield
Sheffield
United Kingdom

ISSN 2192-6255 ISSN 2192-6263 (electronic)
ISBN 978-3-642-45087-7 ISBN 978-3-642-45088-4 (eBook)
DOI 10.1007/978-3-642-45088-4
Springer Heidelberg New York Dordrecht London

Library of Congress Control Number: 2013953860

Printed on acid-free paper

Springer is part of Springer Science+Business Media (www.springer.com)

Acknowledgements

This book is a culmination of doctoral and postdoctoral work in the field of technology ethics completed at the University of East Anglia's School of Environmental Sciences between 2004 and 2011, and the Sustainability Research Institute at the University of Leeds. This research was principally funded by the United Kingdom's Economic and Social Research Council, with additional funding provided by former radioactive waste management organisation UK Nirex Ltd, the Nuclear Decommissioning Authority, and latterly the International Socio-Technical Challenges for Geological Disposal (InSOTEC) research project funded by the Seventh Euratom Framework Programme (FP7).

Thanks go to Peter Simmons and Alan Bond at the University of East Anglia, Elizabeth Atherton of Radioactive Waste Management Limited, and Nick Pidgeon of Cardiff University. I also wish to thank the participants in my workshops, without whom these ethical tools would never have been developed.

Contents

List of Abbreviations

ANT	Actor-Network Theory
BSE	Bovine spongiform encephalopathy
CJD	Creutzfeldt-Jakob disease
COPUS	Committee on Public Understanding of Science
CORWM	Committee on Radioactive Waste Management
DM	Deliberative Mapping
EG	Ethical Grid
EM	Ethical Matrix
HLW	High level waste
IAEA	International Atomic Energy Agency
ILW	Intermediate Level Waste
LLW	Low level waste
MADA	Multi-attribute decision analysis
MCDA	Multi-criteria decision analysis
MODA	Multi-objective decision analysis
MRWS	Managing Radioactive Waste Safely
NEA	Nuclear Energy Agency
NDA	Nuclear Decommissioning Authority
NGO	Non-Governmental Organisations
NIREX	Nuclear Industry Radioactive Waste Management Executive
OECD	Organisation for Economic Cooperation and Development
PSE	Public and Stakeholder Engagement
PTA	Participatory Technology Assessment
PUS	Public Understanding of Science
RAND	Research and Development Corporation
REM	Reflective Ethical Mapping
RWM	Radioactive waste management
SECT	Socially and ethically contentious technology
SM	Stakeholder Mapping
SSSK	Social Studies of Scientific Knowledge
STS	Science and Technology Studies
TA	Technology Assessment
UK	United Kingdom
UNECE	United Nations Economic Commission for Europe
VLLW	Very low level waste

Chapter 1
Risk and Public Involvement in Technology Governance

1.1 Introduction

Certain forms of technology stimulate public controversy when facts about their risks and benefits are unclear, the values underpinning their governance are uncertain, and when emergent ethical issues are difficult to resolve. Numerous examples of technological controversy have surfaced in recent decades: notably genetically modified and synthetic biological organisms, nanotechnologies, stem cell research, renewed nuclear power development, and geo-engineering strategies to mitigate long term anthropogenic climate change. Though each of these examples is fraught with unique social and ethical challenges, what they share is a capacity to generate risks, adversely affecting public welfare and damaging the natural environment. The governance of technological risk involves inevitable trade-offs between different actors, institutions and ecological systems, stimulating academic and practitioner interest in the *social control* of technology in order to establish legitimate and socially robust technology policies. The principal concern of this book is a category of technologies that are grouped under the category label of "socially and ethically contentious technologies" (SECT), defined through their capacity to provoke political controversy, stimulate social movements of opposition, and generate media and academic commentary over their governance.

In the face of social and ethical challenges it has become necessary for governments and technology development organisations to seek to generate support for new and emergent technologies, not just from established elected members, scientific governing bodies, or shareholders; but also from locally affected communities, public interest groups, non-governmental organisations and myriad other 'stakeholders'. Such support is often deemed necessary both as a process of justifying technology policy openly within civil society, and also as a means to defuse the types of public opposition that can result in development failure and wasted public sector and industry resources. As a result, the democratic governance and decision-making processes in science and technology policy within economically developed nations have undergone something of a shift

M. Cotton, *Ethics and Technology Assessment: A Participatory Approach*,
Studies in Applied Philosophy, Epistemology and Rational Ethics 13,
DOI: 10.1007/978-3-642-45088-4_1, © Springer-Verlag Berlin Heidelberg 2014

towards a 'public and stakeholder engagement' (PSE) model, and this has since become a critical concern for academic social scientists and policy practitioners, with a rapid expansion of tools and methods to improve two way communication around the development of technical systems. This introductory chapter explores some of the factors involved in this shift towards participatory technology governance, with particular attention to concepts of risk, science communication and deliberative democracy.

1.2 Socially and Ethically Contentious Technologies

The blanket concept of socially and ethically contentious technologies (SECT) allows us to make general observations about risk-bearing technology projects. The 'technology' aspect requires some elaboration. In many instances, technology can be understood in the familiar sense of referring to the activities of making and using engineered technical artefacts. Such artefacts emerge from the application of scientific and engineering knowledge, which are then packaged in such a form to assist human actors in achieving specific goals and performing particular sets of social functions. In the policy rhetoric surrounding technological development, the role of technology-as-artefact involves a process of socially constructing technology in wholly material terms. Thus defining 'good technology' involves emphasis upon the design quality of the artefact in question: whether it achieves its functions more efficiently than other competing artefacts.

There are of course factors other than efficiency that influence the social desirability of technological artefacts, and these are explored throughout this book. However, it must be stressed that even where social and environmental factors do come into the assessment of technology, these factors tend to be construed as calculable and quantifiable phenomena - risks, costs, environmental impacts, supply chains and price structures. It is as Sørensen (2004) states that "the prevailing image of technology and its designers is as cold as steel, as instrumental as calculation". Processes of governance are grounded in an understanding of design, manufacture, use and disposal of artefacts.

Philosophers and sociologists of technology have long been concerned with broadening the instrumental, material and calculative modes of technological governance; countering the reductionism implicit in construing technology as the physical end product of design, development and manufacturing processes and their associated risks, costs and benefits. A broader taxonomy of the technology concept is thus required, and Mitcham (1994) provides a useful and more varied conceptualisation distinguished by four related dimensions. The first dimension is technology construed as physical, material artefact. Secondly, however, it can be imagined as a form of technical knowledge and also the act of applying knowledge and expertise to the building of such artefacts. Thirdly, it can be a processes or activity, in the sense that technologies are used and applied in certain social contexts, rather than simply existing independently of social action. Fourthly, and most critically in the context of this book, technology can be understood as a social force, a form of volition, in the sense that these artefacts embody cultural, social and moral values, reflect our desires, aspirations and influence our behaviours.

The concept of volition is of critical concern. The development and application of technology is a process that holds both social and technical elements. Artefacts do not stand alone as unconnected objects in an otherwise separate social world. Two important terms have emerged. The first is that of co-production, and the second is that of socio-technical systems. The former describes how technical and social structures are intertwined and co-constitutive - they emerge cyclically and symbiotically with one another shaping and redefining social, moral and technical boundaries. The latter describes complex technological systems in relation to the political, ethical and environmental 'impacts' embedded within them whereby technology development involves heterogeneous associations of human and non-human elements. This is true of many types of technology, and STS scholars have laid bare the complex inter-relationships between technical artefacts and social systems in great detail; whether examining something as geopolitically significant as the Trident Missile Programme (Mort 2001) or as commonplace as indoor plumbing (Sclove 1995) or the bicycle (Bijker 1995). In certain cases, the socio-technical nature of specific technologies (including not just the artefacts themselves but their design, governance, knowledge structures, user interaction, disposal and environment impact) generates a particular kind of collective adverse reaction within the public sphere characterised by adversarial parliamentary debate, the development of opposition actions from civil society stakeholders, local protest groups and non-governmental organisations. In defining what constitutes a SECT in comparison to other technological systems, we must pay attention to the emergent public scepticism, mistrust and organised opposition not only towards the artefacts in question, but also towards the centralised authorities and technical experts charged with the governance of scientific and knowledge and the management of technological risks.

Numerous examples of SECT cases emerge in the STS literature. Recent examples include: the trials of genetically modified organisms (May 1999; Frewer et al. 2004), the renewal of nuclear power (Blowers 2010) and the management of radioactive wastes (Atherton and Poole 2001; Cotton 2012; Blowers and Sundqvist 2010), the disposal of oil rigs (Huxham and D. 1999), the development of nano-technologies (Sheetz et al. 2005), synthetic biology (Schmidt et al. 2009), or Xenotransplantation (Bach and Ivinson 2002). What distinguishes these technologies as socially and ethically contentious is that the vehement public reaction has served to not only undermine the implementation of Government and private sector proposals, but has also catalysed wider public distrust of scientific authority and its relationship to civil society. The reasons for such vehement public reactions are intrinsically linked to the ethics that permeate the development and implementation of these technologies, and in this book I aim to elucidate the relationships between stakeholder conceptions of technological systems and their respective understanding and analysis of the ethical issues that emerge.

1.3 Risk and the Public Acceptance of Technologies

The concept of *risk* is integral to understanding this process of how citizens engage with SECT and with the organisations that govern and implement them. Technology is innately a source of risk because technological artefacts often produce unintended consequences. A fast changing technological environment introduces additional risk because of the numerous social, economic, and political opportunities it creates, and because of the threats it facilitates to human and non-human health and wellbeing (Orman 2013). Accelerating technological change therefore creates compound risks as multiple technological systems interact.

One example referred to throughout this book is that of nuclear fission technologies. The discovery of radioactivity in the early 20th century led to the subsequent development of nuclear fission as a controlled technological process used in war and in domestic electricity generation. The international history of nuclear technologies reveals the potentially catastrophic implications for international relations; dangerous accidents such as the 1986 Chernobyl incident or the more recent crisis in 2011 at the Fukushima Daiichi nuclear reactor in Japan; evidence of cancer clusters around facilities; and the social stigmatisation of communities hosting nuclear technologies, where social, psychological and economic pressures stemming from the perception of risks, cause stress and anxiety amongst the local citizenry. Though inherently dangerous due to the massive energy yields nuclear fission creates, nuclear science also provides low carbon electricity generation (thus potentially mitigating the threat of anthropogenic climate change), has led to radiotherapy in cancer treatment, carbon-dating and a host of other positive practical uses. These emergent positive and negative social and environmental effects mean that the science and technology in question cannot be considered morally neutral. Technologies such as those associated with nuclear science have radically and conspicuously influenced the social sphere, and public understanding of and reactions to new nuclear technologies are deeply entangled in social relationships, shared histories, political power and implicit ethical values. If one were to try and build a new nuclear power station it would become necessary to develop political sensitivity to these powerful socio-technical elements. Untangling these socio-technical relationships is an important aspect of justifying SECTs openly within civil society, and plays a key role in fostering public acceptance of proposals and trust between public actors and institutions.

1.3.1 Defining Technological Risk

The risk phenomenon has been extensively and divergently theorised in the past thirty years, and multiple definitions have emerged in the scientific, social scientific and popular 'lay' understanding of the concept (Renn 1998). To simplify, however, we can understand technological risks as being implicitly related to hazards: the products, processes and other external conditions that threaten the safety and wellbeing of individuals, social groups and non-human entities. Hazards have been often been categorised as external or environmental, such as in the case of earthquakes, droughts or floods; or else anthropogenic, i.e. resulting from human actions such as the generation of radioactive wastes, oil spills or airborne pollutants. In understanding the concept of risk it is important to note that it is through human involvement that events or objects that threaten human or non-human safety are transformed into hazards; for instance, a rockslide or flood on an uninhabited island would not be considered hazardous, rather it would be considered a natural event. Moreover, as our understanding of how human actions alter the structure of environmental systems and their properties, such as in the case of greenhouse gas-induced climate change, the neat epistemological distinction of hazards into 'natural and external', and 'anthropogenic', dissolves. In light of this dialectical relationship between the two, the concept of risk is often considered by social scientists to be the collective contingent effects of anthropogenic technological and developmental processes that generate hazards, and hence human actions and social values are integral to understanding how risks can be identified, calculated and consequently managed and mitigated.

There is significant divergence in the technical definitions of risk within engineering and the mathematical sciences (see in particular Fischhoff et al. 1984); though in general, the technical processes of managing risks have often tended to involve quantifying the likelihood of possible outcomes that result from human decisions, and thus there has been a tendency to rely upon statistical modelling to derive the information upon which risk management practices are framed. In statistical terms, risk exists when known probabilities can be assigned to the outcomes of human decisions. A simple technical definition of risk therefore is as the "product of the probability and consequences (magnitude and severity) of an adverse event (i.e. a hazard)" (Bradbury 1989), leading to risk calculations following a simple metric:

$$Risk = Threat \times Vulnerability \times Impact$$

In this equation *Threat* is the frequency of a hazard; *Vulnerability* is the likelihood of success of a particular threat category against a particular group, individual or organisation; and *Impact* is the total cost of a particular threat experienced by a vulnerable target. In essence all human activities can fit within this rubric of risk assessment. More importantly, efforts to avoid particular risks can generate countervailing risks, which may be of greater probability or magnitude and therefore be more dangerous to human and non-human actors. Risks are therefore managed or minimised rather than totally eliminated. By adopting this type of approach, assessing and mitigating risks involves a focus upon questions of how well risks can be calculated, the level of seriousness that they pose, the accuracy of the underlying science and the inclusiveness of the causal or predictive models used to understand why risks occur and why people respond to them in certain ways (Lupton 1999).

Though risk is often framed in the technical literature as an independently calculable phenomenon, following Beck's (1992) Risk Society thesis, the cultural theory of Douglas (1986), Wildavsky and Dake (1990); and the psychometric approach of Slovic (Slovic 1987, 1993) and others; the concept has been reformulated as a form of social and cultural organisation, stimulating a sustained social scientific critique. Risk is now established as a complex multi-dimensional psychological construct and a form of social discourse. Reframing risk as a complex and multi-dimensional social construct involves paying attention to the wider context of individuals' beliefs, attitudes, perceptions, judgements and feelings, alongside significant questions of ethics and political governance. As a consequence, social scientists and philosophers of risk have consistently called for the incorporation of broader cultural, social and ethical values into the process of risk analysis and management. New methodological tools have emerged to expand upon the technical, statistical definitions of risk, and to produce qualitatively richer and more socially robust risk management practices that incorporate the attitudes, perceptions and values of a great range of affected actors including 'the public'.

This reframing of risk signals a significant narrative shift around new technologies from unproblematic and progressive moral goods, towards fears of disempowerment, ill health and lack of control (Jasanoff 1999; Wynne 2005); hence there has been a sustained academic concern with establishing greater social control of emergent risks. Risk analysis has been augmented into analytic-deliberative risk governance, and STS scholarship has generated new processes of assessment that incorporate both techno-scientific and 'lay' expertise (Renn 1999). Within this (now well established) risk paradigm, researchers have commonly advocated a more inclusive, participatory and open examination of risk cultures, values, and perceptions amongst heterogeneous public and stakeholder actors. Analytic-deliberative approaches to risk governance have emerged under the rubric that public controversies emerging in technological development and implementation can be anticipated and the politics contained (Macnaghten and Chilvers 2012) when governments and private institutions engage with citizens through mechanisms aimed at building trust, dialogue and democratic accountability in decision-making. It is important, therefore, to question who 'the public' are with respect to analytic-deliberative risk governance, in order to better understand the means through which technical authorities and citizens engage in analytic-deliberative risk governance practices.

1.4 The Concept of the Public, and Public Understanding of Science

It is common in the development of new technological artefacts for designers to do user-based research, consumer surveys or focus group activities to better ensure consumer-friendly design and to achieve market competitiveness in relation to rival products. Consumer research into technological design categorically differs from the processes of involving the public in the design and implementation of risk-bearing technologies, however. The notion of SECT implies that there are potentially negative consequences for public welfare, rights or interests that affect those outside of the customer, client or shareholder relationships within the market economy. Economists commonly refer to these effects as *externalities* – examples of where the market fails to internalise negative social and environmental impacts, and so the market price does not reflect the true cost to public social and environmental welfare. In certain circumstances, debate over the management of externalities creates localised controversy between the industry and its immediate neighbours. Examples of this might include the development of onshore wind farms creating visual intrusion and industrialisation of rural landscapes, or the pollution from heavy industry contaminating drinking water supplies. In extreme cases these controversies spill from local and regional concerns into national and, in some cases, international political concerns. There have been a number of high profile health and environmental disasters which have adversely affected public trust in the institutional governance of risk. These include the Bhopal tragedy of 1984, when 500,000 people in India were exposed to deadly methyl isocyanate gas, the 1986 Chernobyl nuclear disaster that spread nuclear contamination across Western Europe, and the Bovine Spongiform Encephalopathy (BSE) or 'mad cow' disease exposure in the United Kingdom linked to variante Creutzfeldt-Jakob disease (CJD) – a lethal form of dementia. Each of these examples exacerbated public scepticism of the moral neutrality of business interests, of organisational risk communication, and the safeguarding of public welfare. This in turn has influenced the ways in which citizens engage with new technologies, as their values and attitudes are filtered through their understanding of previous failures of institutional control.

In understanding why controversy emerges around particular instances of technological risk (and not in others) and why the phenomenon of PSE has arisen to try and solve such problems the concept of *the public* is of critical concern. It is important to understand what if anything constitute this group as a meaningful social entity, and how conceptions of heterogeneous publics by technical authorities inform processes of engagement with non-specialist citizens. This is because developing an understanding of how technical specialists construct, or *imagine* heterogeneous publics is pivotal to understanding the patterns of communication and engagement that emerge (Burningham et al. 2007; Maranta et al. 2003; Walker et al. 2010).

Empirical sociological research into how technical authorities such as planners, scientists and engineers conceive of lay publics reveals a number of common categorisations or social representations of citizen actors. In some cases citizens are conflated with consumers or customers. In others they are 'neighbours' to controversial facilities, are 'voters', self-interested Not-In-My-Backyard (NIMBY) protestors concerned with simply protecting their local turf against unwanted developments (such as new energy technologies like wind farms or nuclear power stations), or else are 'the man/woman on the street' – who is generally construed as uninterested, unaffected and uninformed. In relation to their understanding of and attitudes towards science and technology, citizens are frequently construed as passive; as being (wilfully) ignorant about scientific facts; as polarised for or against technological developments; demanding of zero-risk scenarios; unable to take a strategic viewpoint; incapable of understanding sophisticated technical information; basing their opposition upon non-scientific, soft ethical or political factors; or else as simply the malleable victims of a distorting and sensationalist media that views science as a foundationless, relativistic enterprise (Marris et al. 2001; Burningham et al. 2006; Burningham et al. 2007; Cotton and Devine-Wright 2012; Wynne 1985; Joss and Durant 1995).

These social representations of the public are almost universally negative. Constructing the public as a homogeneous group that is either ill-equipped to be involved in science and generally behaving as a hindrance rather than as valid stakeholders with specific interests in the outcomes and governance of technology development, is a persistent challenge to a socially and democratically robust policy process. Together, these dominant discourses about the attitudes and behaviours of public actors are characterised by what Wynne (1982) terms *deficit model* assumptions about public understanding of science and technology. The deficit model is linked to notions of scientific and technical literacy, i.e. the capacities of non-specialist citizens to understand scientific and technical matters in the manner in which it is communicated by experts. Deficit model thinking has remained a pervasive factor in the management of technological risks, and it construes the public as being opposed to scientific and technological advances due to an inadequate knowledge base – and therefore fundamentally misunderstanding the environmental, social and economic benefits and burdens involved. This in turn leads to a technology policy process that is negotiated within the bounded rationality and objective assessments of privileged scientific and technical experts. Such rationality is defined by an understanding of hard evidence such as costs, safety and environmental performance, which is inevitably prioritised over soft, subjective and consequently irrational public values, feelings and sentiments.

The deficit model assumption of a scientifically illiterate and homogeneous public was the driving force behind the Public Understanding of Science (PUS) movement that dominated government thinking about science-civil society relationships in a number of developed nations including the United Kingdom. For example, UK PUS was institutionalised through the Committee on Public Understanding of Science (COPUS), founded in 1985 by the British Association for the Advancement of Science, the Royal Institution and the Royal Society. The aim of COPUS was to interpret scientific advances and make them more accessible to non-scientists. PUS was largely driven by the findings of large-scale representative national survey studies that tested citizen responses to questions about scientific terms and mechanisms, and these often revealed low levels of *scientific literacy* amongst the citizenry in the US and Europe (Miller 1998). To continue with the nuclear example, PUS surveys in the US have shown that 1 in 10 adults have what could be considered a scientifically correct understanding of the concept of radiation. When asked an open-ended question to explain the meaning of radiation, approximately 11% of respondents provided information that involved the emission of energy as particles or waves. 10% were able to mention the effect that radiation had, but were unable to name a source or explanation of the meaning of radiation (Miller 2004). This type of study is important in understanding why the deficit model has emerged, as these findings appear to have generated an assumption amongst regulatory officials that by somehow developing an adequate mechanism for filling this knowledge gap about how radiation works, and how risks are calculated, that this will encourage 'the public' to adopt a cultural attitude towards science, technology and risk that mirrors that of the technical experts. Thus, the early PUS movement was concerned with finding means to transmit information in a unidirectional manner from expert to lay citizen through processes of simplification and public education.

The underlying assumption inherent to deficit model thinking is that if the public has more knowledge then this will automatically lead to a more positive attitude towards controversial scientific and technological programmes. Early risk communication efforts faced strong criticism as risk communicators tended to present exaggerated claims about the pros of adopting new technology verses the negative consequences of failing to do so as a strategic ploy to help resolve risk conflicts. Nelkin (2002) also notes that in such disputes public and stakeholder concerns can often become reduced to technical questions to be answered by recourse to "better" science alone; if any residual fears still remain it is through information conveyed via risk communication that a resolution is to be found (see also Wardman 2008). However, the communication of information about scientific facts, physical mechanisms, risks, costs and benefits can never be a solely intellectual process that happens in one direction. In part this is because science is commonly understood by citizens to be a process involving disagreements between members of scientific communities, disagreements that are often trans-scientific rather than simply scientific, (for example defining whether genetically modified foods are safe to eat, or for the environments in which they are grown). Public understanding of and support for technology is therefore often embedded in the identities and the social networks of trust from which they emerge. This issue of trust is compounded by a public lack of confidence in scientists' ability to often diagnose the relevant risks accurately. In fact there is a growing concern by citizens that risks, costs and benefits of new technologies may not be well understood so there is little reason to trust the experts at all (Kasperson 1992; Kunreuther 2001).

The former UK Government Chief Scientific Advisor Sir Richard May (1999) called this problem the *patina of distrust*. The patina of distrust refers to how the public may not fully grasp all the scientific complexities of new technologies but are nevertheless aware of the commercial imperatives, sceptical about politics and distrustful of the competence and impartiality of independent regulatory frameworks. The type and severity of resultant public reactions to SECTs are therefore largely dependent upon the level of trust that citizens hold in the institutions (both private industry and governmental) involved. Such trust is closely tied to public confidence in the safety of the technologies and the institutions that put them into practice. Publics must have confidence in governing institutions, which means that they must not only comply with existing legislation regarding safety and regulatory control, but must be also be able to build public confidence through transparency, truthfulness and democratic accountability.

1.5   Technocracy, Democracy and Technology Development

Given the sustained criticism of the PUS model, managing the expert/lay citizen relationship has involved the reformation of science and technology policy around a PSE model involving a pluralistic, two-way communication between technical and non-technical actors. This turn towards multi-party deliberation over technology shifts decision-making processes over design and implementation away from both technocratic and representative democratic forms. Technocratic decision-making emphasises the legitimacy of experts, whereas representative democratic modes emphasise the role of elected officials and aggregate voting systems (through mechanisms such as referenda). In a number of significant cases these two models have been replaced by more participatory and discursive democratic approach, in instances where citizens conceive of the legitimacy of technological proposals in terms of opportunities for their direct involvement. This participatory-deliberative 'turn' to some extent results in increased opportunities for citizens to be involved in policy-making, and so encourages institutions to adapt to new ways of developing technologies within the public sphere. However, despite becoming 'institutionalised' within a number of public and private sector organisations, the underlying rationale for implementing engagement varies widely, depending upon the technology in question and the decision-making context in which it is framed.

The various academic literatures in Geography, STS and Political Science have highlighted a vast array of rationales and motivations for involving different social actors in technology decisions. These have been commonly grouped into three categories of strategic, ethical/normative or substantive rationales (Fiorino 1990). Strategic advantages are the benefits to implementing organisations, such as reducing costs, project delays or public opposition, by resolving stakeholder conflicts and restoring trust in political institutions (Bloomfield et al. 2001) and thus rendering decision-making processes and resultant policies as democratically legitimate (Beierle and Koninsky 2000; Cohen 1989). Ethical motivations emerge from numerous appeals to include public actors in decision-making by invoking concepts of procedural justice, fairness and human rights (NRC 1996; Fiorino 1990; Bohman 2000). This is because engagement practices can present opportunities to examine how new SECT proposals affect the welfare of the citizens that bear localised environmental or health risks, social inequalities and costs when these burdens are concentrated on particular groups, communities or ecosystems and benefits are widely dispersed across broader society. This asymmetrical distribution of goods and bads from certain SECTs can be due to geographical, economic or cultural differences and inequalities, or in some cases geographical and temporal horizons of risk (radiation from nuclear power facilities or climate change resulting from fossil fuel use will affect specific populations both now and in the future). Substantive motivations are when engagement practice is perceived to improve the quality of decisions, making them more socially robust (Beierle 1999). This robustness comes from broader deliberation amongst lay experts which can potentially reveal new kinds of relevant information on social, geographical or moral issues that may otherwise be overlooked in a purely technical analysis. Other potential substantive advantages are for those that actually take part, as deliberation between scientific actors and citizens has been shown to allow opportunities for improvement of the moral and intellectual qualities of the participants (Fearon 1998), and to encourage them to undergo reflexive social learning about technical, social and ethical issues (Tuler 1998).

Irrespective of the underlying rationale, the practice of engagement has become increasingly common within the democratic governance of technology and of the natural environment. The drivers for this have largely stemmed from actors within non-government organisation (NGO), academic and policy circles that have recognised the potential value of deliberation to a healthy democratic society. Two-way collaborative engagement in its various forms is considered by some in academic and policy circles to be a kind of 'gold standard' for decision-making (Felt and Fochler 2008); and this has led to a number of key actors within these academic and policy circles to champion the cause of public engagement as an inherently good or fair thing to practice. Hence, there has been a recent expansion in the literatures on both the development and assessment of new decision-making structures, methods and tools to enable involvement of a broad range of social actors including citizens, and a concerted move to encourage 'upstream' public engagement (Wilsdon and Willis 2004), whereby citizen voices are incorporated into the early design and development phases of new technologies before they become stabilised or 'black boxed' (Latour 2004) at the point of implementation in the public sphere.

Though there are those that see PSE as an inherently good thing due to the emphasis on decentralised power structures and civic empowerment, to some critics public participation in technology decision-making leads to control by public sentiment, leading to the detriment of scientifically defined safety. Such an argument is characteristic of a discourse of *administrative rationalism*, whereby the role of the expert is placed in primacy in social problem solving, and where social relationships of hierarchy are stressed over those of equality or competition (Dryzek 1997). Those that worry over technology policy becoming sentimental are likely to perceive a rise in the prominence of cultural relativism. Scientists and other technical specialists have often been wary of a cultural shift that undermines the foundations of scientific methods and practices, fearing that science is becoming increasingly sidelined by media interests, politicians and in the public imagination. To the engineer as well, the key issues in any given technology may be safety, design efficiency, cost and environmental performance. From this perspective, public acceptability is potentially dangerous because it distracts from the objective of building a carefully engineered solution to maximise safety margins. If science and engineering-based criteria come under attack, this would be detrimental to the success (and safety) of any given design and so, perhaps ironically, citizen involvement would not be in the public interest. To other critics of deliberative decision-making, the intangibility of social and moral factors and the fickle and abstract nature of the public make the incorporation of such values into concrete technologies too difficult to achieve in practice. This factor, combined with the often substantial resources costs involved in designing and evaluating engagement activities, means that institutions in both the public and private spheres may well feel reluctant to innovate in deliberative public-focussed decision-making.

1.6 Technology Assessment

The role of democratic governments in technology policy is also contentious. Governments tend to engage with technology in two ways, which present contradictory demands. Firstly, they promote science and technology through funding and other incentives in order to exploit their social and economic benefits. Proponent of new and emerging technologies, including scientists and other technical actors, expect governmental support in the implementation of new technologies, hoping that public institutions will foster a positive climate for development leading to public acceptance of proposals. Secondly, however, governments are responsible for the regulation of the application of technologies to avoid unintended negative consequences for the citizenry. Citizens expect risks to be controlled, and transparent regulatory mechanisms put in place to ensure their welfare. Given that technologies are developed without citizen oversight, public actors rightfully claim some input into decision-making. In order to reconcile these different tasks, public policy depends on external expertise. The dominant narrative of how science and policy interact is that scientists provide politicians with impartial, factual knowledge in order for partisan politicians to make decision-making in situations of uncertainty. This is commonly referred to as *speaking truth to power*. However, in risk governance, *certainty* is precisely what science cannot provide, and the question of 'how safe is safe enough?' cannot be answered factually by science (all from EUROPTA 2000).

Given the socio-technical nature of SECTs, judgements over what is safe or acceptable involve implicit social, moral and aesthetic values. In response to this, there are those that see the requirement for transparency, democratic accountability and the incorporation of diverse public values as vital interests that must be protected from being overruled by the unidirectional input from a community of scientific experts. As Denenberg (1974) insists: "safety is too important to be left to the experts. It is an issue that should be resolved from the point of view of the public interest, which requires a broader perspective than the tunnel-visioned technicians." The trans-scientific nature of public risk debates mean that scientific data alone is inadequate when assessing something as sociologically complex as the *acceptability* of any given technology.

In response to this, an academic and policy movement has emerged under the rubric of *Technology Assessment* (hereafter TA). The concepts and practices of TA are intended to enhance societal understanding of the broad implications of science and technology and, thereby, to improve the quality and efficacy of political deliberation in fields ranging from environmental management, science policy and military decision-making. Early TA traditionally adopted the aforementioned approach of speaking truth to power, with early adopters trying to gain advance knowledge of technology options, their impacts and consequences, and hence provide and early warning system to encourage governments to steer clear of potential future technological hazards, or else to minimise their harmful effects on society (Decker and Ladikas 2004). There has been a tendency within the TA community to adopt a pragmatic view of technological progress, in that, though TA as a policy movement seeks to shape and evaluate the consequences of technologies it also avoids anti-technology stances, realising that future societal progress is dependent upon technological advancement. Its principle concern is with the ambivalence of technology; assessing the consequences, both and good and bad from implementing particular policies that advance certain technical agendas. The TA motto according to Mohr (1999) is that a new technology must be qualitatively better than the preceding technology, otherwise we do not need it. The concept of *better* differs from that alluded to earlier, as it incorporates socio-economic, ethical and environmental dimensions rather than just quantitative evaluation.

TA in practice has emerged under the auspices of certain civil society bodies in democratic countries. In the United States for example, the U.S. Congress set a global institutional precedent by creating the (now defunct) Office of Technology Assessment (OTA) in 1972. Similar models followed in Europe such as the UK Parliamentary Office of Science and Technology (POST), the Swiss Centre for Technology Assessment, the Office of Technology Assessment at the German Parliament (TAB), the Danish Board of Technology (DBT) and the Belgian Institute of Society and Technology (IST). Each of these bodies was constructed to fulfil an advisory capacity to governments, adopting multi-disciplinary approaches to the analysis and solving of existing societal problems caused by the uncritical application and commercialisation of new technologies, improving the communication practices of scientific research to civil service organisations and stimulating public debate. Though tasked with finding qualitatively better technological solutions that are socially responsive, this necessitates evaluative tools or mechanisms of social and ethical appraisal for new technology developments, and the practices of formalised TA have stimulated growth in this field. Though early TA was dominated by 'hard' evaluative tools such as forecasting, risk analysis, safety assessments, and cost-benefit analyses, in recent years the TA toolbox of methods has expanded to include participatory, dialogic and communicative methods such as consensus conferences, stakeholder workshops, foresight activities and backcasting techniques (Decker and Ladikas 2004; Durant 1999). The task of TA has, like risk analysis, morphed into a more participatory-deliberative structure, with greater opportunities for the involvement of a range of public and stakeholder actors[1].

---

[1] Please note that throughout the book I use the term participatory Technology Assessment (PTA) to describe this model of TA practice, though it must be noted that various 'flavours' of such dialogic TA exist, including Constructive (Rip et al. 1995), Discursive (EUROPTA 2000) and Real-time' (Guston and Sarewitz 2002) models. What each of these share are varying degrees of deliberative citizen involvement and control over technological development programmes. Given that the focus is upon this citizen involvement aspect, I use the PTA as a catchall term to encompass these deliberative-democratic modes of TA.

1.6.1 Technocratic or Participatory-Deliberative Decisions?

The foregoing discussion reveals a clear tension between technocratic and deliberative-democratic paradigms of decision-making when applied to the management of SECTs. Technocracy is embedded in a prevailing discourse of scientific optimism and ecological modernisation that stresses technical rather than social solutions to social and environmental problems; clashing with the new wave of deliberative science, stressing the role of civic expertise, incorporating transparency, accountability and participation. The tension can be summarised as a conflict between those that express concern over weakening the quality of primarily technical decisions by sacrificing scientific accuracy in favour of political expediency, in contrast to those that seek to support the protection of heterogeneous publics' rights to control their own safety, wellbeing and environmental quality. To the latter, the goal of engagement is to provide defence against the indifference and exclusion resulting from technocratic processes, as technocracy is perceived as a form of political oppression, in the sense that it fosters centralised authority structures at the expense of the smaller units of government in which direct participation is possible (Bäckstrand 2004; Fischer 1993; Stirling 2001). Participatory Technology Assessment (PTA) aims to fulfil this promise, to create inclusionary, bottom-up and citizen-focussed oversight of technology governance in the public sphere.

The issue of centralised power over concerned and affected publics is the core political issue. Proponents of PTA insist that the power to make decisions must be placed as far as possible in the hands of the persons who are the most directly influenced by the decision concerned and not in the hands of individual decision-makers and their associated experts. Though proponents of PTA do not underestimate the role of scientific knowledge (especially in regard to defining what is safe and what is not), the inclusion of participatory critiques of scientific knowledge can be justified largely on pragmatic grounds. Technological systems that influence the natural, physical and social environments are frequently complex; by their nature they involve deep systemic uncertainties and so appropriate methods to explore and facilitate a plurality of legitimate perspectives are necessary. This contrasts with the practice of science which, broadly speaking, generates a picture of reality designed for controlled experimentation and abstract theory building. The physicist-turned-philosopher of science Thomas Kuhn (1962) showed the inadequacy of scientific thinking to solve complex, multi-dimensional problems. He suggested that scientific inquiry normally consists of puzzle solving within a largely unquestionable framework or paradigm. Puzzle-solving or *normal* science can be very effective with complex phenomena reduced to their simple atomic elements. However, it is not best suited for the tasks of complex decision-making over implementation of technological solutions, or policies which affect social and ecological environments. This is because scientists are primarily trained with an eye to 'the technical agenda of science' (Funtowicz and Ravetz 1993), whereby the practical upshot of theoretical knowledge is the central focus. Broadly speaking, the scientific mindset fosters expectations of regularity, simplicity and certainty in the phenomena and in our interventions, but these can inhibit the growth of our understanding of complex problems and of appropriate methods to their solution. Hence, the methodologies of normal laboratory science are of restricted effectiveness in advising optimal courses of action in the context of new and emerging risks to our society (whether from anthropogenic climate change, nano-technologies etc.) in part because conventional, normal science actually involves very little re-thinking of what scientific knowledge means to society and what actually counts as legitimate expertise (Wynne 1996).

PTA offers lay expertise as a means to help in resolving complex technical problems. This is most clearly illustrated in Wynne's widely cited study of sheep farming in Northwest England following the Chernobyl accident, where the advice of experts to minimise livestock exposure to the irradiated environment failed to take into account local knowledge of the landscape or the expertise of local farmers, leading to unnecessary widespread contamination and destruction of agricultural produce and deepened public distrust of scientific experts (Wynne 1996). Though non-scientific expertise is important to technology policy, developing a satisfactory relationship between the two is by no means a simple process. There are cases when the public do not feel (or do not want to feel) qualified to make well informed decisions and take responsibility for action; in other cases they believe that they are the experts. Thus, the expert/lay person relationship is contingent upon the context of the individual, the situation and the knowledge under consideration.

1.6.2 Post-normal Science

As a potential solution to the challenging nature of these dialogues between the scientific and lay experts, Funtowicz (1993) and Ravetz (1999) postulate the idea of incorporating *post-normal science* into decision-making processes. The post-normal moniker relates to the Kuhnian normal scientific paradigm, and applies to issues where facts are uncertain, values in dispute and the stakes are high. Post-normal science and technology policy, "involves going beyond traditional assumptions that science is both certain and value-free, it makes system certainties and decision stakes the essential elements of its analysis" (Ravetz 1999). In normal science day-to-day scientific research practice involves review of results within standard peer communities of other experts. In some cases other stakeholders have access to and availability to critique this knowledge. These include professional consultancy organisations, that take the knowledge gained from available (usually peer-reviewed) science and either apply it to well-characterised problems in the context of policy and decision-making, or else contribute to knowledge dissemination in the so-called *grey literature*. Post-normal science goes beyond both of these practices, and involves understanding the nature of highly uncertain, publicly contested knowledge which occurs in the context of many health, safety, and environmental decisions (the GMO example is one of particular relevance).

The quality of post-normal science cannot be assured by standard peer-review processes because of the uncertainty and social and moral values involved. Instead, proponents suggest that work of this nature be subjected to extended peer review, involving not only scientists but also a much broader range of stakeholders affected by the use of science. In essence, the concept of *accountability* is key, rather than simply the quality control of peer review. A post-normal technology decision-making process would move beyond the sole use of technical tools (for example probabilistic risk assessments) to a method where the quality of the process of research, planning and implementation is paramount. This involves enabling joint learning and joint planning between technologists, users and those affected. This ensures a grounding of science and technology within the social context in which it is applied or discussed. The role of experiential knowledge is therefore elevated to a similar status as scientific knowledge. Citizens become part of extended *peer communities* (Ravetz 1999) providing alternative information and critique. The inclusivity of this type of approach means that public engagement is intrinsically linked to citizenship (Ackerman and Fishkin 2003; Mendelberg 2002; Barber 1984, 1998). The lay citizenry contribute to the evaluation of technology problems, in part because it serves their own purposes as self-interested consumers or as stakeholders representing specific interest groups in the outcome of a decision; but their involvement is often also motivated by a desire to represent their community, their society or their environment, and thus goes beyond personal interests. The idea of the *scientific citizen* is therefore an emergent concept in the practice of PTA.

1.7 A Critical Evaluation of the Deliberative Turn

So far I have asserted that the problems of technology governance essentially involve a divide between technocratic and deliberative-democratic approaches. However, this is a rather simplistic means of looking at the problem of expertise and the nature of democratic legitimacy in technology planning. First of all it is important to note that there is no uniform, generally applicable model of public engagement that can be applied in all circumstances, and to all technology problems. This vagueness surrounding the methods and design of participatory approaches often results in industry and governmental bodies applying inappropriate methods to decision-making processes, or else trying to persuade and inform under the pretence of stimulating public deliberation. In essence there is a risk of dressing up public relations to look like public engagement. This happens because the language of 'engagement' is often used to disguise processes in which public support is manipulated into a pre-chosen proposal by an elected (or financially wealthy) decision-makers, and hence participatory rhetoric belies a smokescreen for achieving pre-defined ends (see for example Hindmarsh and Matthews 2008). Participatory processes can therefore become political tools through which citizens become co-opted into formalising top-down, authority-made decisions, thus providing a veneer of social legitimacy to what could otherwise be considered technocratic decision-making (Allen 1998; Chess and Purcell 1999; Gariepy 1991). It is necessary therefore to gauge the extent to which participatory processes devolve power to publics, and so evaluation criteria such as Arnstein's ladder of participation (1969) are necessary. The ladder approach shows progressive levels of democratic involvement from manipulation and information provision at the lowest levels, moving through consultation towards, partnership, delegated power and then citizen control at the upper levels. The ladder shows the varying levels at which public actors have influence in technology policy, with those that emphasise decisional influence of citizens towards the top, and those that foster centralised power towards the bottom. Though widely cited, this ladder approach misses out a significant temporal dimension of participation. Governing organisations may have the will to install citizens in a top-level partnership or citizen control model of decision-making. However, if this occurs too late in the decision time-line to effectively facilitate public deliberation and identify alternatives, then decision-makers may fall back on uncritically accepted dominant or default options. This would result in a loss of faith in participation from both the perspectives of the implementing organisations and the participants. Governing organisations may be less willing to invest in future participatory forums in light of previous failures, and so getting it right must work first time. This means that governing organisations must take care to manage participant expectations, ensuring that what is really information transmission is not hidden in the language of deliberation, as this would serve not only to undermine those that take part and the objectives of the exercise, but it also damages the legitimacy and credibility of the decision-making process itself and consequently trust in the institutions involved.

Even timely and well organised participation may not necessarily improve the substantive quality of decision-making, as it may serve to exacerbate rather than quell controversy, leading parties to deconstruct one another's positions instead of deliberating effectively. There are, therefore, a number of significant pitfalls involved in simply replacing technocratic with deliberative-participatory decision-making. As Collins and Evans (2002) argue, the participatory-deliberative turn has replaced the problem of legitimacy (i.e. from reliance on expert opinion), with a problem of extension whereby the involvement of many different voices in participatory procedures can be a hindrance to effective decision-making. Fischhoff's (1995) reflective essay on the development of risk management practice, notes an evolution of ideas that tracks in a similar way to Arnstein's ladder, evolving through progressive stages to become less centralised and technocratic towards being more citizen-centred and participatory-deliberative over time. Early risk management operated on a model of technological decision-making that involved technical actors just "getting the numbers right", moving towards processes of communicating those numbers, and then trying to explain them. This evolved into processes of convincing people that they've accepted similar risks in the past, and then "being nice to them", followed by a period in which the goal was: "all we have to do is make them partners". Though risk management has shown a progressive democratic shift over the 20 year period that

Fischhoff covers, the institutions that instigate such participatory risk management processes have continuously failed to reflect on whether or not they provide the best quality information, are inclusive of a range of stakeholder actors or were effective in creating policy change. It is important to note that despite the positive rhetoric of academic and policy actors favouring PSE, simply running workshops or consensus conferences does not automatically translate into better decisions, and some participatory methods may simply be considered too small or *ad hoc*, unrepresentative of key segments of civil society or too issue-specific to have a significant influence on the governance of science and technology policy. To be effective, Fischoff *et al*. and Stern and Fineberg (1996) suggest that a successful process must involve good quality and relevant science, the right participants and suitable participatory processes, and so develop an accurate, balanced and informative synthesis of these heterogeneous elements in order to create an effective analytic-deliberative process.

The issue of inclusivity and representativeness is crucial under these circumstances, as engagement practices can have, perhaps ironically, significant exclusionary effects. They may tend to bias the viewpoints of individuals that have the resources (e.g. enough free time) and the motivation to participate. There are inevitably those who are at risk of being marginalised by policy decisions and yet suffer the greatest barriers to taking part in deliberation - for example, the time constrained working poor, non-native language speakers, or those that struggle to gain access due to illness or disability. These groups (amongst others) will often have the greatest stake in the outcomes of policy proceedings and yet may have least access. Participation under these circumstances can lead to policy outcomes that widen the gap between those that are able (and willing) to use these opportunities to be involved in decision-making and those that are not (Mansbridge 1980; Young 2000), and will hence act contrary to the egalitarian ideals of pluralistic democratic involvement enshrined in the PTA movement. Moreover, participatory processes though aiming to capture the voices of lay publics, will often become subject to influence from interest groups such as activist organisations, NGOs and businesses (Bartlett and Baber 1989). In such cases the public arena of deliberation becomes a site for activism, lobbying or grandstanding, rather than open dialogue.

The solutions to these problems are contingent upon two important prerequisites. Firstly, it is necessary to encourage an accurate representation of affected publics, and thus avoid the biases of class, gender, race and sexuality-based distinctions within participatory processes. This can largely be solved through effective sampling measures. Methods such as Deliberative Polling (Fishkin et al. 2000; Fishkin 1995) involve a randomly sampled microcosm of a chosen population, thus ensuring a pattern of participation that includes representation from those voices often excluded from decisions. Such a forum would mean providing resources, time, funding and access to information in order for engagement to be balanced, meaningful, and allow participants to remain autonomous and involved throughout. The second prerequisite is the establishment of decisional influence. Both the process and outcomes of deliberative methods must be shown to have a positive, practical and demonstrable effect upon technology implementation, otherwise participation serves to exacerbate public opposition, mistrust and civic disengagement, rather than promote active citizenship, as it places citizens in the position of reacting to predefined proposals rather than providing input to their development. When this occurs, citizen actors are likely to seek alternative opportunities to halt the development process in line with their wishes, whether through exerting political pressure upon locally elected representatives, or else engaging in organised protest or direct actions to halt the development deemed by them to be unethical.

1.8 Conclusions

PTA is by no means the panacea to resolve every social and ethical controversy that surrounds nascent technological programmes. However, the potential advantages to individuals, institutions and civil society make deliberation a persuasive form of political governance despite the potential drawbacks. As Gutmann (1993) and Johnson (1998) suggest, deliberation involves reasoned and critical discussion rather than presumed cultural consensus, technical authority or political deal-making, and so it is superior to aggregative, voting-based or technocratic decision-making in achieving the goal of democratic legitimacy, representation and fair outcomes for affected segments of civil society. The direct inclusion of individuals in the political and ethical discussion of technology implementation remains important because the implicit consent involved in technocratic decision-making or national and regional voting (in electoral politics and representative forms of democratic process) is insufficient to legitimately expose individuals to additional or elevated risks, costs and other burdens that may result without informed consent. Inclusive participation is required so that consent can be obtained explicitly and transparently from those affected, improving the procedural fairness of all manner of decision-making processes and hence improving the democratic validity of a range of possible policy outcomes. However, though PTA provides opportunities for the social control of technology, this does not automatically equate with establishing the ethical legitimacy of technologies, and this issue is discussed in the following chapter.

References

Ackerman, B., Fishkin, J.S.: Deliberation Day. In: Fishkin, J.S., Laslett, P. (eds.) Debating Deliberative Democracy. Blackwell, London (2003)

Allen, P.T.: Public Participation in Resolving Environmental Disputes and the Problem of Representativeness. Risk: Health, Safety and Environment 9, 297–308 (1998)

Arnstein, S.R.: A ladder of citizen participation. Journal of the American Institute of Planners 35(4), 216–224 (1969)

Atherton, E., Poole, M.: The Problem of the UK's Radioactive Waste: What Have We Learnt? Interdisciplinary Science Reviews 26, 296–302 (2001)

Bach, F.H., Ivinson, A.J.: A shrewd and ethical approach to xenotransplantation. Trends in Biotechnology 20(3), 129–131 (2002), doi:10.1016/s0167-7799(02)01917-0

Bäckstrand, K.: Scientisation vs. Civic Expertise in Environmental Governance: Eco-feminist, Eco-Modern and Post-modern Responses. Environmental Politics 13(4), 695–714 (2004)

Barber, B.: Strong Democracy: Participation Politics for a New Age. University of California Press, Berkley (1984)

Barber, B.: A passion for democracy. Princeton University Press, Princeton (1998)

Bartlett, R.Y., Baber, W.E.: Bureaucracy or analysis: implications of impact assessment for public administration. In: Bartlett, R.V. (ed.) Policy Through Impact Assessment: Institutionalized Analysis as a Policy Strategy. Greenwood Press, Westport (1989)

Beck, U.: Risk Society: Towards a New Modernity. Sage, London (1992)

Beierle, T.C.: Using social goals to evaluate public participation in environmental decisions. Policy Studies Journal 3(4), 75–103 (1999)

Beierle, T.J., Koninsky, D.M.: Values, conflict, and trust in participatory environmental planning. Journal of Policy Analysis and Management 19(4), 587–602 (2000)

Bijker, W.E.: Of Bicycles, Bakelites and Bulbs: Toward a Theory of Sociotechnical Change. The MIT Press, Cambridge (1995)

Bloomfield, D., Collins, K., Fry, C., Munton, R.: Deliberation and inclusion: vehicles for increasing trust in UK public governance? Environment and Planning C: Government and Policy 19(4), 501–513 (2001)

Blowers, A.: Why dump on us? Power, pragmatism and the periphery in the siting of new nuclear reactors in the UK. Journal of Integrative Environmental Sciences 7(3), 157–173 (2010)

Blowers, A., Sundqvist, G.: Radioactive waste management - technocratic dominance in an age of participation. Journal of Integrative Environmental Sciences 7(3), 149–155 (2010)

Bohman, J.: Public Deliberation: Pluralism, Complexity and Democracy. MIT Press, Cambridge (2000)

Bradbury, J.: The policy implications of differing concepts of risk. Science, Technology & Human Values 14(4), 380–399 (1989)

Burningham, K., Barnett, J., Carr, A., Clift, R., Wehrmeyer, W.: Industrial constructions of publics and public knowledge: a qualitative investigation of practice in the UK chemicals industry. Public Understanding of Science 16(1), 23–43 (2007)

Burningham, K., Barnett, J., Thrush, D.: The limitations of the NIMBY concept for understanding public engagement with renewable energy technologies: a literature review. Manchester University, Manchester (2006)

Chess, C., Purcell, K.: Public Participation and the Environment: Do We Know What Works? Environmental Science and Technology 33(16), 2685–2692 (1999)

Cohen, J.: 1989. Deliberation and democratic legitimacy. In: Hamlin, A., Pettit, P. (eds.) The Good Polity: Normative Analysis of the State. Blackwell, Oxford (1989)

Collins, H.M., Evans, R.: The third wave of science studies: Studies of expertise and experience. Social Studies of Science 32, 23–296 (2002)

Cotton, M.: Industry and stakeholder perspectives on the social and ethical aspects of radioactive waste management options. Journal of Transdisciplinary Environmental Studies 11(1), 8–26 (2012)

Cotton, M., Devine-Wright, P.: Making electricity networks 'visible': industry actor representations of 'publics' and public engagement in infrastructure planning. Public Understanding of Science 21(2), 17–35 (2012)

Decker, M., Ladikas, M.: Bridges Between Science, Society and Policy: Technology Assesment - Methods and Impacts. Springer, Berlin (2004)

Denenberg, H.S.: Nuclear power: Uninsurable. Congressional Record: US Government Printing Office, Washington DC (1974)

Douglas, M.: Risk Acceptability According to the Social Sciences. Sage, London (1986)

Dryzek, J.S.: The Politics of the Earth: Environmental Discourses. Oxford University Press, Oxford (1997)

Durant, J.: Participatory technology assessment and the democratic model of the public understanding of science. Science and Public Policy 26(5), 313–319 (1999)

EUROPTA, European Participatory Technology Assessment: Participatory Methods in Technology Assessment and Technology Decision-Making. Copenhagen: The Danish Board of Technology (2000)

Fearon, J.D.: Deliberation as discussion. In: Elster, J. (ed.) Deliberative Democracy. Cambridge University Press, Cambridge (1998)

Felt, U., Fochler, M.: The bottom-up meanings of the concept of public participation in science and technology. Science and Public Policy 35(7), 489–499 (2008)

Fiorino, D.: Citizen Participation and Environmental Risk: a Survey of Insitutional Mechanisms. Science, Technology & Human Values 15(2), 226–243 (1990)

Fischer, F.: Citizen participation and the democratization of policy expertise: From theoretical inquiry to practical cases. Policy Sciences 26, 165–187 (1993)

Fischhoff, B.: Risk perception and communication unplugged: 20 years of process. Risk Analysis 15(2), 137–146 (1995)

Fischhoff, B., Watson, S.R., Hope, C.: Defining Risk. Policy Sciences 17, 123–139 (1984)

Fishkin, J.: The Voice of the People. Yale University Press, New Haven (1995)

Fishkin, J.S., Luskin, R.C., Jowell, R.: Deliberative polling and public consultation. National Centre for Social Research, London 53(4) (2000)

Frewer, L., Lassen, J., Kettlitz, B., Scholderer, J., Beekman, V., Berdal, K.G.: Societal Aspects of Genetically Modified Foods. Food and Chemical Toxicology 42, 1181–1193 (2004)

Funtowicz, S., Ravetz, J.: Science for the post-normal age. Futures 25(7), 739–755 (1993)

Gariepy, M.: Toward a dual-influence system: Assessing the effects of public participation in environmental impact assessment for hydro-Quebec projects. Environmental Impact Assessment Review 11(4), 353–374 (1991)

Guston, D.H., Sarewitz, D.: Real-time technology assessment. Technology in Society 24, 93–109 (2002)

Gutmann, A.: The Challenge of Multiculturalism in Political Ethics. Philosophy and Public Affairs 22, 171–206 (1993)

Hans Mohr, H.: Technology Assessment in Theory and Practice. Techné: Journal of the Society for Philosophy and Technology 4(4) (1999)

Hindmarsh, R., Matthews, C.: Deliberative Speak at the Turbine Face: Community Engagement, Wind Farms, and Renewable Energy Transitions, in Australia. Journal of Environmental Policy & Planning 10(3), 217–232 (2008)

Huxham, M., Sumner, D.: Emotion, science and rationality: The case of Brent Spar. Environmental Values 8(3), 349–368 (1999)

Jasanoff, S.: The Songlines of Risk. Environmental Values 8(2), 135–152 (1999)

Johnson, J.: Arguing for Deliberation: Some Skeptical Considerations. In: Elster, J. (ed.) Deliberative Democracy. Cambridge University Press, Cambridge (1998)

Joss, S., Durant, J.: Public Participation in Science. Science Museum and European Commission Directorate General XII, London (1995)

Kasperson, R.E., Golding, D., Tuler, S.: Siting Hazardous Facilities and Communicating Risks Under Conditions of High Social Distrust. Journal of Social Issues 48, 161–167 (1992)

Kuhn, T.S.: The Structure of Scientific Revolutions. University of Chicago Press, Chicago (1962)

Kunreuther, H., et al.: Risk, Media and Stigma: Understanding Challenges to Modern Science and Technology. Earthscan (2001)

Latour, B.: Politics of Nature: How to Bring the Sciences into Democracy. Harvard University Press, Cambridge (2004)

Lupton, D.: Risk and Sociocultural Theory: New Directions and Perspectives. Cambridge University Press, Cambridge (1999)

Macnaghten, P.M., Chilvers, J.: Governing risky technologies. In: Lane, S., Klauser, F., Kearnes, M. (eds.) Critical Risk Research: Practices, Politics and Ethics. Wiley Blackwell, London (2012)

Mansbridge, J.: Beyond adversary democracy. Basic Books, New York (1980)

Maranta, A., Guggenheim, M., Gisler, P., Pohl, C.: The Reality of Experts and the Imagined Lay Person. Acta Sociologica 46(2), 150–165 (2003)

Marris, C., Wynne, B., Simmons, P., Weldon, S.: Public Perceptions of Agricultural Biotechnologies in Europe Final Report. University of Lancaster, Lancaster (2001)

May, R.: Genetically modified foods: faults, worries, policies and public confidence. In: Note by the UK Chief Scientific Adviser, Office of Science and Technology, London (1999)

Mendelberg, T.: The Deliberative Citizen Theory and Evidence. In: Political Decision Making, Deliberation and Participation, pp. 151–193 (2002)

Miller, J.D.: Public Understanding of, and Attitudes Toward Scientific Research: What We Know and What We Need to Know. Public Understanding of Science 13(3), 273–294 (2004)

Miller, J.D.: The measurement of civic scientific literacy. Public Understanding of Science 7(3), 203–223 (1998), doi:10.1088/0963-6625/7/3/001

Mitcham, C.: Thinking Through Technology: The Path between Engineering and Philosophy. University of Chicago Press, Chicago (1994)

Mort, M.: Building the Trident Network: A Study of the Enrollment of People, Knowledge, and Machines. The MIT Press, Cambridge (2001)

Nelkin, D.: Communicating risk: Once again. Political Communication 19(4), 461–463 (2002)

NRC. Understanding Risk: Informing Decisions in a Democratic Society. National Research Council: National Academy Press, Washinton DC (1996)

Orman, L.: Technology as Risk. IEEE Technology and Society Magazine 32(2), 22–31 (2013)

Ravetz, J.R.: What is Post-Normal Science? Futures 31, 647–653 (1999)

Renn, O.: Three decades of risk research: Accomplishments and new challenges. Journal of Risk Research 1(1), 49–71 (1998)

Renn, O.: A model for an analytic-deliberative process in risk management. Environmental Science and Technology 33(18), 3049–3055 (1999)

Rip, A., Schot, J.W., Misa, T.J.: Constructive Technology Assessment: A New Paradigm for Managing Technology in Society. In: Rip, A., Schot, J.W., Misa, T.J. (eds.) Managing Technology in Society. The Approach of Constructive Technology Assessment, pp. 1–12. Pinter Publishers, New York (1995)

Schmidt, M., Ganguli-Mitra, A., Torgersen, H., Kelle, A., Deplazes, A., Biller-Andorno, N.: A priority paper for the societal and ethical aspects of synthetic biology. Systems and Synthetic Biology 3(1-4), 3–7 (2009), doi:10.1007/s11693-009-9034-7

Sclove, R.: Democracy and Technology. Guilford Publications, London (1995)

Sheetz, T., Vidal, J., Pearson, T.D., Lozano, K.: Nanotechnology: Awareness and Societal Concerns. Technology in Society 27(3), 329–345 (2005), doi:10.1016/j.techsoc.2005.04.010

Slovic, P.: Perception of risk. Science 236(4799), 280–285 (1987)

Slovic, P.: Perceived risk, trust and democracy. Risk Analysis 13, 675–682 (1993)

Sørensen, K.H.: Cultural Politics of Technology: Combining Critical and Constructive Interventions. Science, Technology & Human Values 29(2), 184–190 (2004)

Stern, P.C., Fineberg, H.V.: Understanding Risk: Informing Decisions in a Democratic Society. National Academy Press, Washington DC (1996)

Stirling, A.: Participatory processes and scientific expertise: precaution, diversity and transparency in the governance of risk. Participatory Learning and Action 40, 66–71 (2001)

Tuler, S.: Learning Through Participation. Human Ecology Review 5(1), 58–60 (1998)

Walker, G., Cass, N., Burningham, K., Barnett, J.: Renewable energy and sociotechnical change: imagined subjectivities of 'the public' and their implications. Environment and Planning A 42(4), 931–947 (2010)

Wardman, J.: The Constitution of Risk Communication in Advanced Liberal Societies. Risk Analysis 28(6), 1619–1637 (2008)

Wildavsky, A., Dake, K.: Theories of Risk Perception: Who Fears What and Why? Journal of American Academy of Arts and Sciences 119(4), 41–60 (1990)

Wilsdon, J., Willis, R.: See-through Science: Why public engagement needs to move upstream. Demos, London (2004)

Wynne, B.: Rationality and Ritual: The Windscale Inquiry and Nuclear Decisions in Britain. The British Society for the History of Science, Bucks (1982)

Wynne, B.: From Public Perception of Risk to Technology as Cultural Process. In: Covello, V., et al. (eds.) Environmental Impact Assessment Technology and Risk Analysis. Springer, Berlin (1985)

Wynne, B.: May the Sheep Safely Graze? A Reflexive View of the Expert-Lay Knowledge Divide. In: Lash, S., Szerszynski, B., Wynne, B. (eds.) Risk, Environment and Modernity. Sage Publications, London (1996)

Wynne, B.: Risk as globalizing discourse? Framing subjects and citizens. In: Wynne, B., Leach, M., Scoones, I. (eds.) Science and Citizens: Globalization and the Challenge of Engagement. Zed Books, London (2005)

Young, I.M.: Inclusion and democracy. Oxford University Press, Oxford (2000)

Chapter 2
Ethics and Technology

2.1 Introduction

In the previous chapter deliberative public engagement towards technology policy
was presented as a necessary means to achieve democratically legitimate and
socially robust outcomes when risks, costs and other social and environmental
benefits and burdens are distributed asymmetrically between social groups and
ecological systems. The arguments within this book are grounded in a normative
ethical commitment to deliberative democratic control of technology governance
despite the various drawbacks associated with representativeness and legitimacy
discussed in chapter 1. The grounding assumption is that pluralistic involvement
of heterogeneous publics in participatory Technology Assessment (PTA) can
assure that decisions are substantively fairer than those that are based upon
technical expertise alone. Public trust in institutions gained through fair and open
decision-making may help to foster broader acceptance of controversial
technology proposals when they would have been otherwise objected to. At the
very least, the processes and methods of PTA legitimise public objections, in the
sense that they are transparent to decision-makers and based upon a process of
justification through open deliberation. This procedural fairness aspect of
decision-making has been shown under certain circumstances to alleviate public
scepticism towards implementing institutions and to build public support for
decisions taken, even when they are politically unpopular amongst an affected
community (Gross 2007; Renn et al. 1996). However, though it provides
opportunities to enhance democratic legitimacy, this should not be confused with
an assessment of the ethics of technology policy and practice, as these two facets
are ontologically distinct.

I begin this chapter by aiming to delineate the concept of the ethical legitimacy
of technology from the political legitimacy that stems from broad public support.
This is important because these two interrelated facets are often discussed side by
side in popular discourses about the public acceptability of socially and ethically
contentious technologies (SECT). Ethical issues have often risen to the forefront
of public debates about technological control (Paula 2001); and care must be taken
not to conflate one with the other. To illustrate this point, I return to the nuclear

power example. Continuous public debate over new nuclear build has shown fluctuations in support influenced by the outcomes of global events. In particular, the 2011 Fukushima Daiichi nuclear power plant crisis in Japan caused a number of European nations including Germany, Italy and the United Kingdom, to reassess the validity of their nuclear new build programmes, regardless of whether or not such proposals were likely to be affected by similar environmental conditions that caused damage in Japan (factors such as regional seismic activity, flooding and institutional control). Political will to continue with the expansion of nuclear power in Europe has been heavily influenced by public concerns over safety in the wake of this catastrophe. Risk is culturally mediated and so individuals anchor their understanding of the concept of technological safety through pre-existing biases, heuristics and social representations. Whether or not it is right or fair to implement new build nuclear power on the basis of these culturally mediated values and perceptions, is itself a significant ethical issue. However, perceptions of fairness are not the same as actual fairness. Deciding what is fair involves attention to a variety of factors, such as how certain locations will bear greater risks than others, the overall benefits to society from low carbon electricity versus the harms from nuclear catastrophe or leaking radioactive waste repositories, or the meta-ethics of decision-making processes which include or exclude certain voices. It also raises an important distinction between different forms of values that citizen actors may hold. Questions over the ethical legitimacy of SECTs stand independently of the strength of public concerns over safety, as intensifying public fear does not equate to the technologies becoming more dangerous.

Continuing a policy of new build nuclear power may be a less popular political decision in the years following the 2011 Japanese nuclear crisis, but assessing the technological desirability of nuclear power based solely upon public favour would reduce justification to *argumentum ad populum* - an appeal to the popularity of a decision, though this alone is insufficient to suggest that a technological proposal is ethically justified. We are presented with the *Is-Ought* distinction (Hume 1739), questioning how to relate a descriptive analysis of the values and judgements held by citizen actors based upon prevailing views about trust, safety, and perceptions of fairness; with a prescriptive normative ethical assessment of the actions and consequences of its implementation.

2.2 The Challenge of Technology Ethics

Just as we must not conflate normative ethics with descriptive ethics, similarly, when describing the different kinds of values that people hold towards technological developments it is important to distinguish ethical principles with other sorts of expressed values. In all cases of SECT, public judgements about acceptability embody a range of aesthetic, cultural, religious and political values as well as ethical ones. To give an example, Genetically Modified Organism (GMO) controversies in the UK were frequently framed in media discourse in terms of 'Frankenstein Foods', involving the 'tampering with nature' or 'playing God'. Such responses are frequently portrayed in the emergent media and academic commentary as reflecting particular sorts of public morals, though in many respects such discourses reflect positions more akin to aesthetic values. They reveal distaste and concern for hubris in scientific advancement, and suggest that modern society should maintain the integrity of a construct called *Nature*. Though such values are extremely important in the wider discussion of the acceptability of SECT, a value system that posits GMOs as creating a Frankenstein food does not mean that such biological research processes are *de facto* ethically unjustified. Just because something is unnatural does not automatically make it wrong. This conflation of ethical and aesthetic axiologies with an instinct of revulsion is termed a 'yuck factor' by bioethicist Arthur Caplan (1994). Schmidt (2008) illustrates this point by asking us to imagine living in a drought-stricken area, to be told by an engineering firm that from now on your drinking water would come from recycled sewage. Though the concept of reclaimed sewage has the potential to ensure long-term safe water resource usage, and ease pressure on aquifers and other limited water supplies, the first reaction to the proposal might be to feel a sense of repugnance. It is undoubtedly an instinctive reaction. Rejecting fearsome or repugnant things, especially when those things are unfamiliar, has been an important part of our evolutionary development. What Schmidt suggests is that if this yuck factor is shared by large groups of (voting) people then the collective repugnance can fuel a social force with the power to shape public policy making in ways that are not always desirable. Yucky reactions may be valid, useful even, for the individual. They provide an important warning sign for gauging the social acceptability of technological proposals. It is not necessary or desirable to brand those that find technologies repugnant as irrational, or Luddites (or whatever other name calling often gets used in these debates). Nor should we attempt to exclude them from public discourse. What I suggest, however, is that the ethical assessment of the technology must not be swayed entirely by such rhetoric; in other words decision-making over what is right in technology and environmental policy must not devolve into moral panic without room for independent philosophical justification.

Yucky feelings about technology are complex in their make-up because they mix together ethical judgements, instincts and other forms of social, cultural, religious and aesthetic values and norms. Separating or at least identifying these different facets is a significant problem for the assessment of SECT and unravelling this problem requires philosophical guidance. This has proved challenging, in part because of a general paucity of philosophical perspectives within the field of Technology Assessment. Though a vast array of perspectives on the political and cultural acceptability of specific technological risks has emerged within the social sciences, perhaps surprisingly, there is comparatively little research from the traditions of normative ethics. Academics within this field have largely tended to avoid the discussion of technologies that belong to the domain of engineering (Roeser and Asveld 2008). For social scientists committed to re-examining the values inherent to the design and implementation of technological artefacts, there has been a concerted effort to shift the focus away from a sole examination of instrumental values such as efficiency, safety, utility, reliability, and ease of use, towards examining the substantive values of technology in a social and cultural context. In liberal democracies, such values would likely include liberty, justice, privacy, security, friendship, comfort, trust, autonomy, and transparency (Flanagan et al. 2008). Understanding the nature and justification of these broad values through the lens of normative ethics has, however, proved challenging.

2.3 Technology Blindness in Normative Ethics

It is clear that risk bearing technologies co-evolve with moral thought and action; though this process has rarely been acknowledged, let alone thoroughly understood by moral philosophers. This is because normative ethics as a discipline within moral philosophy has tended to focus upon the role of human actors, their conduct and the consequences of their actions. The term *normative* is often defined as describing the effects of the particular structures of culture that regulate the functions of social activity (Phillips 1979); normative ethics is thus by extension, concerned with those prescriptions and abstract theories that provide shape to the outcome of social activities. In doing so, dominant normative traditions have distinct concepts of what should constitute the moral good, and hence varying conceptions of what is right and wrong. Normative ethics is a social and political philosophy with a practical goal. It prescriptively guides and governs the conduct of human nature. A normative ethical theory can never list right and wrong actions (even if it were a very long list), the theory must obtain some level of abstraction from the particular and a degree of generality in order to successfully deal with differing circumstances and actors in a comprehensive and systematic way. The aim of this branch of ethics is to bring unity to the multifarious judgements, evaluations, rules and principles that exist in society by trying to develop a coherent set of procedures to represent, organise and justify them. The goal is to arrive at a set of moral standards that regulate the conduct of moral actors, which may involve stipulating the habits that one should acquire, the duties that one should follow, or the consequences of behaviour on others.

Normative ethics has, to an extent, been dominated by contractarian theories such as Kantian deontology (with a focus upon the duties of the individual), Benthamite utilitarianism (with a focus upon maximising the benefits to or welfare of the greatest number), or Rawlsian notions of justice (with a focus upon the conditions for redistributing social benefits). Though these normative ethical theories have been richly elaborated in the moral philosophy literature, I avoid a discussion of the relative merits of each in turn, as the focus within this book is upon the meta-ethical conditions of technology assessment, rather than a focus upon a single normative approach. It is important to note, however, that these normative theoretical approaches have, in turn, strongly influenced the disciplines of applied ethics of which technology ethics is a part. Applied ethics is often construed as the process of making normative ethical theories practically useful tools for real-world decision-making. Applied ethics has often involved overlaying specific normative theories on specific situations or fields of knowledge where there is a strong need for ethical guidance and critical evaluation. It is potentially sub-divisible into *special ethics*: investigating areas of human endeavour where moral guidance is needed such as environmental ethics, business ethics, bioethics and so on; and *practical ethics*: concerned with providing tools or techniques to allow practitioners within these fields to make better informed judgements by critically assessing a range of normative positions. Applied ethics has tended to involve doing the philosophical work of ethical assessment within the theoretical and conceptual realm of normative ethics and then applying, or perhaps imposing that theoretical work on practical decision-making. In this way it can be understood in the same way as other applied disciplines such as applied mathematics or biology whereby a pool of theory is applied in a top-down manner to a real-world context.

This is rather simplistic portrait of applied ethics, however. The field has, in recent decades, moved away from the rather staid debates that dichotomise deontological-versus-utilitarian normative analyses, and the inevitable arguments over which should be applied. A range of alternative moral philosophies have emerged that variably focus upon other aspects of ethical decision-making and moral action beyond discussions of duties-versus-consequences. For example there has been a resurgence in theoretical discussion that emphasises the value of practical moral wisdom (Nussbaum 1986); virtues, moral narrative and individual character (MacIntyre 1984), concepts of care and feminist ethics (Kheel 1993; Shrage 1994), or else upon the psychological, subjective, imaginative and metaphorical nature of ethical reasoning (Dunn 2004; Haidt et al. 1993; Kekes 1991; Werhane 2002). Just as I do not wish to outline each individual normative theory, similarly a discussion of the nature of and relationship between normative and applied ethics is also beyond the scope of this book. The meta-ethical challenge, so to speak, is to understand how individuals come to ethical judgements, and how these can be shaped by normative theoretical considerations to reach a philosophically robust evaluation of ethical issues. In doing so, we must examine how ethical judgements are formulated, and how technology as both artefact and social process fits into such judgements. Rather than seeking a mono-theoretical solution, I aim in this chapter to reflect upon what counts as justification in the ethical assessment of technology.

One of the principal meta-ethical points I wish to consider, is that the dominant traditions of normative ethical reasoning, whether deontological, utilitarian, justice or virtue-based have tended to focus upon the moral actor as the centre of analysis. In doing so there is a tendency to then ascribe a neutral role to technologies. These dominant normative ethical traditions have tended to frame technologies as passive objects that are manipulated by moral actors, so normative ethics has commonly focussed upon behaviours and norms adopted by the users and developers of technology, and have had less to say about the ethical status and agency of the artefacts themselves. Early work in the field of technology ethics focussed upon the responsibilities of engineers, scientists and technicians; in particular the openness and transparency of their professional practice. Much of what we understand as technology ethics today had its roots in engineering ethics, one of the specialist ethics disciplines. Engineering ethics has been principally concerned with developing stringent ethical codes of practice for practitioners. Issues such as *whistle-blowing* predominate - whereby engineers take responsibility upon discovery of unethical practices, or the adverse social and environmental consequences of specific technological innovations. More recently, however, this focus on engineering practice has been viewed as inadequate, or at least incomplete. This inadequacy of engineering ethics lies in its incapacity to assess the ethical consequences of technology in full, simply due to the fact that engineers are not the only important actors in technological design, governance, application and use. Indeed technology assessment as a discipline is partly geared towards reducing the moral authority of engineers in shaping the outcomes of technology development processes.

Though the ethical ramifications of engineering professional practice are undoubtedly important, our technologically mediated world requires a more holistic and visionary perspective on ethical mattes. As Johnson and Powers (2005) suggest, the social world is filled with human-made objects, which enable and inhibit human thought and moral action, informing how we think, act, and arrange ourselves into social units and institutions. Technological artefacts, knowledge processes and practices serve to dichotomise the human-made from the natural world, though the two are of course deeply intertwined. The natural world has been so affected by technological development that some such as Allenby (2004), suggest that we now live in the age of the Anthropocene - an increasingly anthropogenic, technologically mediated planet. This is important because our concept of technology ethics cannot remain focussed solely upon the bearers and appliers of technical knowledge – scientists and engineers, because as Mitcham's (1994) model from the previous chapter suggests, technology is not just about artefacts and designers, it is a form of volition, ubiquitous and integral to our way of living and being in the natural and social world. It covers our aspirations, social networks and personal identities. Magnani (2007) extends this point to suggest that humans are increasingly integrated with nonhuman artefacts and technical processes, and are therefore deserving of an entirely new understanding as hybrids or 'things' as a means of according them the proper respect. We must conclude, therefore, that our technology ethics must extend to understanding not only how engineers or technologists behave when faced with artefacts that harm or benefit the social and natural world, but also how they shape and inform social and cultural practices, individual experiences, communities and environments in their development, use and governance. Most importantly we must learn how technologies co-evolve with our social ethics.

2.4 Actor-Network Theory and Technology Ethics – Bridging Disciplinary Divides

In resolving this meta-ethical problem, I turn first to the insights of the allied fields of Social Studies of Scientific Knowledge (SSSK) and Science and

Technology Studies (STS). These disciplines have been instrumental in revealing the power that technology holds within society. As previously mentioned in chapter 1, STS has revealed the co-evolutionary development of technologies within society, whereby not only are cultural and moral values implicitly embedded within technological design practice, but also that with every new technological development the social world shifts, and reacts – moulding our values and practices. This entangled relationship between social actors, values and technological artefacts has been explored by one prominent strand of SSSK called Actor-Network Theory (hereafter ANT).

Actor-Network Theorists, notably Michael Callon and Bruno Latour, have been concerned with the nature of technology within social networks, studying the interdependent social practices that constitute work in science and technology. What differentiates ANT from traditional sociological understandings of social networks is that it views the actors that constitute the network of science and technology as consisting not only people and social groups, but also artefacts, devices, and entities. ANT asserts that social networks are heterogeneous, containing many dissimilar elements, and thus can be understood as socio-technical systems (Latour 1987, 1995; Callon 1987). Ultimately ANT is a theory of semiotics - asserting that entities take form and acquire attributes through relationships with other entities. An ANT analysis involves every aspect of a technology's development, planning, policy, use and disposal, drawing together diverse elements such as building materials, contractors, designers, workers, machinery, environmental systems, even the paper upon which the proposals are written and the blueprints for design. ANT purports to show how all of these artefacts have a generalised symmetry (Latour 1993) – i.e. they must be described in the same terms as social agents within the network, and thus these heterogeneous elements become epistemologically related as *actants*. Actants take shape by virtue of their relations with one another, with no special status given to human agents over animals, or non-human artefacts. Action is a process where each of these elements is caught up in the web of relations. The ontology of ANT is therefore flat structured between material and ideational elements. Under a principle of generalised symmetry the heterogeneous actants have equal footing, thus the theory rejects technological determinism and social determinism as descriptors of technology development and social change (Callon 1987).

ANT emphasises actantiality forged through alliances and negotiations between human and non-human actors, so power is accumulated and maintained through alliances with technologies, materials and other non-human allies, as well as those with other people. Change is understood as a process of stability or instability within relationships between human and non-human elements; thus it is construed as a process of struggle to hold relationships in place. Relations need to be repeatedly "performed" or the network will dissolve, so analytical attention focuses on the ways in which different actants attempt to increase the remit of their actions by holding other (actors and artefacts) in place, and escaping this holding effect that others impinge upon them (Latour 1993). Different actants may be "enrolled" as "allies" to reinforce network relationships and the stability and

form of these actants should be seen as a function of the interaction of
heterogeneous elements as these are shaped and assimilated into a network of
assemblages (Law and Hassard 1999).

Though a full discussion of the strengths and weakness of ANT is beyond the
scope of this book, it is important to note a general flaw in ANT's conception of
technology networks due to the inherent complexity of the task it sets. An ANT
researcher must question the point at which she must stop including new artefacts
or actors in the web of inter-related action; a consideration termed the 'problem of
selection' (Walsham 1997). The decision of actant inclusion or exclusion involves
specific judgements, to avoid an endless circular, descriptive process of network
analysis. As such, ANT has been criticised as unnecessarily 'long-winded',
involving nothing more than a descriptive account of all the involved actants
within a network (ibid). Indeed the problem of the actant relationship is that the
intentionality derived from human reasoning is largely absent from the analysis, as
actions are borne from interactions and alliances of heterogeneous human and
non-human elements. Though generalised symmetry presents an ontologically
controversy for the social sciences, ANT has value in its capacity to explain why
technology implementation, policies or social endeavours succeed or fail, by
paying attention to the changes that occur in the integrity of the networks in which
these elements are embedded (referred to as the Entelechy). Latour (2005) argues
that the crumbling of a network is the reason for failure of a particular technology.
The failure of technology is when programmes for development are halted due to
adverse public reactions, like the 2011 German nuclear policy example, can be
reviewed in relation to success or failure of the networks within such
technological proposals are embedded, and of the networks that implementing
organisations failed to build, extend or stabilise.

ANT is particularly pertinent to the discussion in this book due to its evolving
relationship with moral philosophy. At first glance, it appears that ANT and
normative ethics show significant incompatibilities. As mentioned before, normative
ethics has traditionally focussed upon the individual and the choices that are
available to them. This is problematic to an ANT conception of technology ethics,
because it asserts that technologies and other artefacts have ethical value and
influence within complex actor-networks. An ANT-focussed ethics of technology
would include the analysis of influence that technological artefacts exert within the
social realm. As we have seen, such an approach contrasts with some of the
dominant modes of thinking in normative ethics that conceptualise technology as a
set of tools in the hands of rational moral actors. Studies of technology ethics have
often maintained that technologies are fundamentally ethically neutral because it is
the cognitive process of individual moral judgement that controls how they
are used. This instrumental vision of technology posits them as a means to an end
and so the choice of technological means to solve problems is thus a morally neutral
affair (Van De Poel 2001). Such an instrumental vision of technology as a passive
tool is inadequate because it brackets moral action away from the tools and
resources that enable or inhibit such moral action. We must consider that when
actors formulate specific goals to be met by a technology, that this cannot be

separated from the development and choice of technological means to meet those goals. In other words, technologies are not always developed with clear goals in mind and their development can influence the social and moral choices after they have been realised. For example a number of medical advancements such as Aspirin and Penicillin have changed the way medicine is practiced (in suppressing fever and killing bacteria respectively) but neither was specifically "invented" with those clear consequentialist ethical goals. The development of technologies does not involve a single objective or set of choices, there are always a number of alternatives with differing environmental and social effects, each of which has unforeseen consequences, both desirable and undesirable, so one cannot truly argue that technologies are in any sense value free or ethically neutral. With this in mind, ANT has relevance to ethical assessment of technology, although in many respects, we would be ill-equipped to deal with the moral problems of technology using the language and concepts of ANT alone. When using ANT in the assessment of social ethics there are a number of important obstacles; some theorists such as Walsham (1997) and Bijker (1993) have raised concerns that ANT studies of technology networks show a fundamentally 'amoral' and 'apolitical' stance, encouraging the devaluation of the role of human actors. As neither actor nor artefact is given priority over the other, the so-called 'actantiality' of each is a reflection of the quality that provides actants with their actions, with their subjectivity, their intentionality and with their morality (Latour 1999). Essentially the networked relationships between actants is what gives them moral value. As a result, ANT appears at odds with much of the body of modern moral philosophy.

The principle antagonism between ANT and normative ethics is based in part, upon the language used. ANT tends to use a rather peculiar language to describe the ethical issues of actant relationships: it discusses technologies in terms of their success or failure, winning and losing according to the stability and strength of their respective networks. This language is something of an obstacle when trying to assess which technological artefacts are morally desirable, and similarly there is a paucity of concepts for examining the underlying political choices that influence technology decisions. The focus on winning and losing has led some critics of ANT to accuse theorists of adopting confrontational and even militaristic terminology (Radder 1992) whereby the success of technologies is put in terms of strategies and alliances, and hence allies and opponents; a terminology that has an implicit morality all of its own. In short, ANT has been accused of de-humanising decision-processes by placing technological artefacts on equal footing with human actors, and thus lacking the vocabulary to adequately assess the moral aspects of technology decisions (Radder 1992; Winner 1993). As Radder (1992) and Keulartz et al in particular (2002) recognise, we have an enormous body of ethical theory on the one hand which tends towards simplistic, instrumental views of technology, but a broad and complex conception of human moral action; and on the other, science and technology studies (specifically ANT), that has a complex constructivist conception of technology but a distinct and rather peculiar conceptualisation of ethical values. The ultimate goal of a successful technology ethics is thus to bridge the two disciplines in a manner that retains the practical value of both theory approaches while providing a cohesive and practically useful philosophy of the social ethics of technology decisions. Part of this bridging effort involves a focus on the language and concepts used, and one important project within technology ethics is find the right vocabulary to assess the complex ethics of sociotechnical systems.

2.5 Ethical Theory and Participatory Technology Assessment Practice

The relative ethical neutrality of technology in moral philosophy is not the only challenge facing a deliberative, public-focussed ethical assessment. One must also question the value of applying normative ethical theories to a technology in order to achieve a just outcome for society. At the risk of generalising about normative ethical theory, one must be wary of theoretical monism when using and applying normative ethical theories. Normative monism is the assertion that an ultimate set of guiding principles can be discovered through rational inquiry, one that has practical benefits in freeing societies from prejudice and dogmatism; setting forth comprehensive systems from which to orient one's judgements, carving up the moral landscape so that one can systematically arrange the relevant issues and think more clearly and confidently about moral problems (Pojman 1995). There are those moral philosophers that react against a perceived cultural pluralism and ethical relativism within Western society, suggesting that relativism reveals a loss of confidence in traditional authorities and inherent value of ethical theories in elucidating moral problems. Such proponents of normative theory suggest that a rational approach to ethics is vital if society is to survive and flourish. To the normative monist it is believed that one can then clarify how principles and values relate to one another and crucially offer some guidance on how to live. Though I paint the picture of normative theoretical monism with a broad brush, we can generalise to an extent in saying that monists assert that what is right and wrong does not change between societies, or time frames; in stark contrast to ethical relativism, which highlights the flexibility of ethical systems to change with time, across civilisations and societies, emphasising traits such as transience and reflexivity. Relativism is an expression of the idea that there is no single true doctrine in ethics; there are different views and some may be true for some people, while others are true for other people (Blackburn 2000). If relativism holds that unconditional truth cannot be ascribed to any one ethical position or theory, then it in turn provides support for the notion of pluralism and toleration; if no single belief or set of beliefs is superior to any other in terms of truth then all must be accorded equal respect.

Herein lies the meta-ethical dilemma for PTA. The politics of democratic nations within which PTA practices are embedded, are (at least nominally) culturally pluralistic, seeking to ensure that policy-making remains open to diversity and does not arbitrarily exclude minority positions or marginalised voices. If one were to accept the foundations of such pluralism within society, one must also question whether such pluralism should extend to an acceptance of a broad array of ethical values held by different groups, cultures and communities. The challenge lies in finding a role for normative and evaluative ethics in a culturally diverse world. Though a breadth of perspectives is necessary, one must also be committed to finding some metric or standard against which to measure the validity of those ethical values: a fundamental meta-ethical problem for incorporating prescriptive ethical theories into pluralistic political decisions.

Throughout this book I argue for a comprehensively deliberative approach to assessing the ethics of technology given this problem of pluralism. If we were to base our justification solely upon monistic normative theory then we would tend to appeal to general and universal decision-rules that remain abstract, regardless of the specificities of the case. This is a challenge to participatory-deliberative decision-making, because this mode of thinking is by contrast, case specific, culturally plural and philosophically diverse. Therefore, neither a single ethical theory approach nor a multitudinous set of ethical theory approaches can adequately provide a practical solution to the problems presented, given the universal character of the ethical standards they purport and their competing definitions of the moral good. Hence there is a general incompatibility between the application of specific normative theories and pluralistic deliberative decision-making.

The problem stems from the role of negotiation, consensus building and the pragmatic value of theory in both cases. In participatory-deliberative decision-making the emphasis is on the practical implications of negotiation between and, in some cases, consensus building among participants. Encouraging conflicting and antagonistic groups to accept and validate one another's values and positions involves compromise, mutual learning and negotiation between the involved parties. It is this quality that allows deliberators to reach agreements, or at the very least clarify the terms of their disputes, reaching a 'consensus about dissent' (Raiffa 1994), with the hope of improving the quality of the decisions that are made. Monistic normative ethics is largely incompatible with this approach. In normative ethical justification, negotiation is at the very least undesirable. One of the central elements of a negotiation process involves convincing others to accept the accuracy or plausibility of information that will influence their decision. To a normative ethical theory that is grounded in metaphysics, processes of negotiation are at best inappropriate and at worst, counter-intuitive to the search for objective ethical truth. Moral philosophers have been rather reluctant to rely upon the negotiation skills of individuals, due to this clash with the ontological validity of theoretically coherent maxims or rules.

To put it crudely, if we were simply to apply a normative ethical theory as a straightforward applied 'tool', this would create a 'top-down' ethics, with metaphysics at the top and technological design and governance practice at the bottom. If we were to adopt an ethically relativist approach we would begin with public moral values and judgements and extend these upwards to the evaluation of design and governance. However, both of these approaches are inadequate. A top-down approach would exclude the plurality of perspectives that emerge from public responses to new technologies. It would, in essence, produce another form of technocracy, though this time one of ethical absolutism. Aside from the problems of meta-ethically justifying one theory or corpus of rules over another, it

would also add significant political difficulties to what are likely to be difficult
and protracted decision-making processes. If a philosopher were to wade in to a
debate over a SECT, analyse the 'right' course of action and then apply it without
recourse to public input, then this would likely cause people to react negatively to
the judgement, thus antagonising affected stakeholder actors to the detriment of
the decision-making process.

The counter to this of course is that it would be similarly unacceptable to
simply allow participants in a Technology Assessment process to decide on what
is right or fair, simply on the basis of their own opinions, prejudices or
unconsidered moral judgements; which could be considered reactionary, and
philosophically destitute. Nor could they simply choose from a selection of
theories and decide the most appropriate course of action on the basis of applying
theory to case. A robust ethical TA process must try to bridge this divide, allowing
those involved to both engage with, evaluate, critique and conclude a course of
action from a both a 'bottom-up' perspective, in the sense of engaging with
individual stakeholders' broad array of ethical views and values, and top-down in
the sense of maintaining 'ontological validity' i.e. being grounded in an
understanding of the philosophical conception of ethics, and I return to this issue
in the next chapter.

2.6 Whose Ethics Counts?

Finally within this chapter, there is one other meta-ethical consideration that we
must attend to: one that bridges this to the previous discussion over the
involvement of citizen actors in TA. If we accept that technology ethics is
complex and co-evolves with society, that it must be assessed by more than just
the engineers, and involve more than simply the application of normative ethical
theory, then the next question then becomes, whose ethics do we consider as
important or valid, and who should be in charge of deciding what is right? As I
stated in the previous chapter, it is important to assess the values implicit in
technological development in the context of participatory governance, through a
TA process that incorporates a range of voices and perspectives. In practice,
however, this is not always the case.

When the question of ethics arises in complex socio-technical debates over new
technologies, very often the first response by governing organisations is to
establish an independent council or ethics committee designed to address and
evaluate the problem. This has become common practice, internationally, where
science and technology ethics is deliberated upon within the context of national
and governmental ethics commissions or other forms of institutionalised oversight
bodies, as a means to guide and inform moral practice. This is most prominent in the
field of bioethics; for example, in the United States within the last thirty years, a
variety of bioethics commissions have played an advisory role to the White House
and Congress on health and life science issues. In Europe there are examples of
powerful committees such as the German Ethics Council or the UK's Nuffield

Council on Bioethics, and Siegetsleitner (2011) notes that similar models have emerged in the Developing World (for example Gabon, Ghana, Guinea, Jamaica, Togo and the Republic of El Salvador). In other instances, councils or committees have emerged around specific forms of technology from international bodies such as the standing ethics committee of the Human Genome Organisation (HUGO) and the Nanoethics Advisory Board to and the Food Ethics Council. These groups are commonly composed of experts from diverse academic and professional backgrounds charged with assessing the 'ethical impacts' of proposed technologies such as gene therapies, human cloning, novel foods or nano-technologies.

We must question then from what form of expertise are these councils and advisory bodies composed. Siegetsleitner (2011) continues by stating that most commissions are comprised of experts in the fields of medicine, biology, law, political and social sciences, theology and philosophy. Though medical and biological scientists contribute their medical and scientific expertise in an expert advisory capacity, legal experts propose legal formulations and social scientists can comment upon social values and political context; we must question firstly whether experts who are not philosophers can contribute to the evaluation of ethical issues, and secondly, whether the philosopher can do any better.

2.6.1 The Role of Scientists and Philosophers in Ethical Assessment

In the previous chapter, the role of the technical expert was under scrutiny, alongside the shift in reliance upon scientific and technical expertise towards participatory-deliberative technology policy. In a number of scientifically advanced democratic nations, including the UK, scientific and technical knowledge has lost some of its privileged status. Decisions over the implementation of SECTs, be they GMOs, nanotechnologies or nuclear energy are no longer framed solely in terms of technical criteria and by those that are deemed to have expert judgement. When it comes to discussing issues of ethics, however, 'the scientist' (however this category is defined) has two main roles in public debate. The first is to maintain specific standards of research ethics. Research ethics mainly focuses upon standards of practice in scientific practice (and indeed other forms of social scientific, arts and humanities research). Research ethics covers issues such as protecting research subject's autonomy and welfare in human and animal experimentation, protecting anonymity and scientific protocols to reduce heuristic bias in the reporting of findings, avoiding misconduct through plagiarism and falsification of data, and complying with safety standards and regulatory controls. Like engineering ethics, these standards of ethics are practice-oriented, concerned with maintaining the highest standards of professional conduct. However, scientists also engage with ethics in a second manner. It is often the case that scientific specialists are called upon to explain the mechanics of the scientific processes involved in controversial scientific developments, however, they are also often required to provide ethical commentary on them (Miah 2005). Part of the new public-facing role of the scientist in this era of 'impact' driven scientific research is now to weigh in to key debates on the social and ethical value of scientific findings, and consequently the implications to society, to the economy, and to the natural environment from new technologies that emerge from basic scientific findings. This is inherently problematic from a meta-ethical perspective, as scientific and technical experts often lack specific insight into the ethical implications of the scientific discovery itself. As Turner (2001) states:

> "...if experts are the source of the public's knowledge, and this knowledge is not essentially superior to unaided public opinion, i.e. not genuinely expert, the public itself is presently not merely less competent than the experts but is more or less under the cultural or intellectual control of the experts."

Scientists and engineers certainly possess expertise, but expertise and familiarity with a research topic and its consequences should not be confused with expertise in the application of normative ethical theory, nor in providing robust moral judgements. Given the aforementioned problems of public controversy emerging over technocratic decision-making, it appears that scientific input into ethical assessment can be flawed due to a lack of demonstrable competence in making ethical judgements. In short, scientists are not ethics experts, and if technology policy is significantly shaped by the proscribed moral viewpoints of scientific authorities, then this is, in essence another form of technocracy, one that would likely exacerbate further public conflict. If competence is the issue, then one line of thought suggests that the scientist should simply be replaced by the ethicist. In practice this has sometimes been the case in these ethics advisory bodies. Moral philosophers have been called upon to apply a specifically formulated moral judgement based within theory, in the rather top-down manner alluded to earlier. Such judgement is therefore expected to be philosophically purer or more robust than one which any 'ordinary', non-specialist citizen could provide. This is in essence another type of top-down ethics, but one of expertise rather than ontology.

2.7 Conclusions

Thus far I have referred to the concept of 'top-down' ethical decision-making in describing a process of applying ethical theory perspectives to technology problems in the classical applied manner. Examining the roles of scientists and philosophers within ethics councils and committees does however raise a second instance in which ethics could be considered top down, in the sense of the specialist-centred assessment of technology development and implementation. Ethical assessment that is top-down in the sense that it is primarily based upon specialist input is deeply contentious. The advice of specialists, whether they are professional philosophers or not, is insufficient to ensure a balanced judgement even when there are a selection of viewpoints available (Reber 2006). This is because despite their technical or ethical theory expertise, such 'experts' have no special insight into right and wrong, justice and injustice. As Rawls (1995) argued, there are essentially no experts in moral matters, philosophers have no more moral authority than other citizens. Trained ethicists have no superior competence or knowledge on normative matters to specially qualify them as moral arbiters and their opinions are not qualitatively 'better' than that of the non-expert because trained ethicists have no special access to or monopoly on moral truth. They may possess technical competence, however normative problems are not technical questions (Baylis 2000; Imwinkelried 2005). Therefore one must question whether ethical experts can adequately represent the diversity of moral values and viewpoints that emerge from PTA processes. Given that ethicists are (in the main) adherents to specific normative perspectives, can such experts really speak on behalf of public interests? I contend that that an expert or ethical-specialist centred approach presents a new form of 'ethical technocracy' that mirrors all the previous criticisms of techno-science centred policy-making, and so in fashioning a decision which is both ethically and politically legitimate, we must consider alternative arrangements that place ethical assessment back into the hands of the non-specialist citizenry who are ultimately affected by SECTs. In order to do this though, we must encourage our citizens to consider a range of ethical theory perspectives, reflect upon the judgements that they make, and the cultural, religious and moral biases inherent in those judgements, and then make ethically robust decisions that are attentive to the decision at hand. In the following chapter I outline the meta-ethical groundwork of an approach designed to achieve this aim.

References

Allenby, B.R.: Engineering and Ethics for an Anthropogenic Planet. In: Emerging Technologies and Ethical Issues in Engineering, pp. 7–28. National Academies Press, Washington DC (2004)

Baylis, F.: Expert Testimony by Persons Trained in Ethical Reasoning: The Case of Andrew Sawatzky. Journal of Law, Medicine and Ethics 28, 224–231 (2000)

Bijker, W.E.: Do Not Despair: There Is Life after Constructivism. Science, Technology and Human Values 18(1), 113–138 (1993)

Blackburn, S.: Relativism. In: LaFollette, H. (ed.) Ethical Theory. Blackwell Publishing, Oxford (2000)

Callon, M., Hughes, T.P.: Society in the Making: The Study of Technology as a Tool for Sociological Analysis. In: Bijker, W.E., Hughes, T.P., Pinch, T.J. (eds.) The Social Construction of Technological Systems: New Directions in the Sociology and History of Technology, pp. 83–103. MIT Press, Cambridge (1987)

Caplan, A.: If I were a rich man could I buy a pancreas? Indiana University Press, Bloomington (1994)

Dunn, R.: Moral Psychology and Expressivism. European Journal of Philosophy 12(2), 178–198 (2004)

Flanagan, M., Howe, D., Nissenbaum, H., Weckert, J.: Embodying Values in Technology:
   Theory and Practice. In: van den Hoven, J. (ed.) Information Technology and Moral
   Philosophy. Cambridge University Press, Cambridge (2008)
Gross, C.: Community perspectives of wind energy in Australia: The application of a
   justice and community fairness framework to increase social acceptance. Energy
   Policy 35(5), 2727–2736 (2007), doi:10.1016/j.enpol.2006.12.013
Haidt, J., Koller, S., Dias, M.: Affect, culture, and morality, or is it wrong to eat your dog?
   Journal of Personality and Social Psychology 65, 613–628 (1993)
Hume, D.: Treatise on Human Nature: of Virtue and Vice in General. Oxford University
   Press (1739)
Imwinkelried, E.J.: Expert Testimony by Ethicists: What Should be the Norm? The Journal
   of Law, Medicine & Ethics 33(2), 198–221 (2005)
Johnson, D.G., Powers, T.M.: Ethics and Technology: A Program for Future Research. In:
   Mitcham, C. (ed.) Encyclopedia of Science, Technology, and Ethics, pp. xxvii–xxxv.
   Thompson Gale, Farmington Hills (2005)
Kekes, J.: Moral Imagination, Freedom and the Humanities. American Philosophical
   Quarterly 28(2), 101–111 (1991)
Keulartz, J., Korthals, M., Schermer, M., Swierstra, T.E.: Pragmatist Ethics for a
   Technological Culture. In: The International Library of Environmental, Agricultural
   and Food Ethics, vol. 3. Kluwer, Dodrecht (2002)
Kheel, M.: From Heroic to Holistic Ethics: The Ecofeminist Challenge. In: Gaard, G. (ed.)
   Ecofeminism: Women, Animals, Nature, pp. 243–271. Temple University Press,
   Philadelphia (1993)
Latour, B.: Science in action. How to follow scientists and engineers through society.
   Harvard University Press, London (1987)
Latour, B.: We have never been modern. Harvester Wheatsheaf, Hemel Hempstead (1993)
Latour, B.: La science en action: introduction à la sociologie des sciences (Science in
   action: introduction to the sociology of science). Gallimard, Paris (1995)
Latour, B.: On Recalling ANT. In: Law, J., Hassard, J. (eds.) Actor Network Theory and
   After, pp. 15–25. Blackwell, Oxford (1999)
Latour, B.: Reassembling the Social: An Introduction to Actor-Network-Theory. Oxford
   University Press, Oxford (2005)
Law, J., Hassard, J.: Actor Network Theory and After. Blackwell, Oxford (1999)
MacIntyre, A.: After Virtue: A Study in Moral Theory. University of Notre Dame Press,
   Indiana (1984)
Magnani, L.: Morality in a Technological World: Knowledge as Duty. Cambridge
   University Press, Cambridge (2007)
Miah, A.: Genetics, Cyberspace and Bioethics: Why not a public engagement with ethics?
   Public Understanding of Science 14(4), 409–421 (2005)
Mitcham, C.: Thinking Through Technology: The Path between Engineering and
   Philosophy. University of Chicago Press, Chicago (1994)
Nussbaum, M.: The Fragility of Goodness: Luck and Ethics in Greek Tragedy and
   Philosophy. Cambridge University Press, Cambridge (1986)
Paula, L.: Ethics: the key to public acceptance of biotechnology? Biotechnology and
   Development Monitor 47, 22–23 (2001)
Phillips, D.L.: Equality, justice and rectification: an exploration in normative sociology.
   Academic Press, London (1979)
Pojman, L.P.: Ethics: Discovering Right and Wrong. Wadsworth Publishing Company,
   Belmont (1995)

Radder, H.: Normative reflections on constructivist approaches to science and technology. Social Studies of Science 22(1), 141–173 (1992)

Raiffa, H.: The Art and Science of Negotiation. Cambridge University Press, Cambridge (1994)

Rawls, J.: Reply to Habermas. The Journal of Philosophy 92(3), 132–180 (1995)

Reber, B.: The Ethics of Participatory Technology Assessment. Technikfolgenabshätzung - Theorie und Praxis 2(15), 73–81 (2006)

Renn, O., Webler, T., Kastenholz, H.: Procedural and Substantive Fairness in Landfill Siting: A Swiss Case Study. Risk: Health, Safety & Environment 7(2), 145–168 (1996)

Roeser, S., Asveld, L.: The Ethics of Technological Risk. Earthscan, London (2008)

Schmidt, C.W.: The Yuck Factor When Disgust Meets Discovery. Environmental Health Perspectives 116(12), 524–527 (2008)

Shrage, L.: Interpretative Ethics, Cultural Relativism and Feminist Theory. In: Shrage, L. (ed.) Moral Dilemmas of Feminism, pp. 162–184. Routledge, London (1994)

Siegetsleitner, A.: Ethics in Trouble: A Philosopher's Role in Moral Practice and the Expert Model of National Bioethics Commissions. In: Garner, B., Pavlenko, S., Shaheen, S., Wolanski, A. (eds.) Cultural and Ethical Turns: Interdisciplinary Reflections on Culture, Politics and Ethics, pp. 41–50. Inter-disciplinary Press, Oxford (2011)

Turner, S.: What is the Problem With Experts? Social Studies of Science 31(1), 123–149 (2001)

Van De Poel, I.: Ethics, Technology Assessment and Industry. TA-Datenbank-Nachrichten 2(10), 51–61 (2001)

Walsham, G.: Actor-Network Theory and IS research: Current status and Future Prospects. In: Lee, A.S., Liebenau, J., DeGross, J.I. (eds.) Information Systems and Qualitative Research, pp. 466–480. Chapman and Hall, London (1997a)

Werhane, P.H.: Moral Imagination and Systems Thinking. Journal of Business Ethics 38, 33–42 (2002)

Winner, L.: Upon Opening the Black Box and Finding It Empty: Social Constructivism and the Philosophy of Technology. Science, Technology & Human Values 18(3), 362–378 (1993)

Chapter 3
Pragmatism, Public Deliberation and Technology Ethics

3.1 Introduction

In the previous chapter I asserted that the participatory assessment of socially and ethically contentious technologies (SECT) must pay attention to three meta-ethical considerations. Firstly, that a technology ethics must pay attention to the influential role of technological artefacts in shaping social moral values, and in inhibiting and enabling the moral actions of individuals embedded within complex actor-networks. Secondly, that the application of normative ethical theories in a classical metaphysics-down-to-practical matters way is insufficient to ensure a balanced range of judgements that reflect the broad plurality of moral perspectives present within society. And thirdly, that the judgements of experts, be they scientists or moral philosophers is contested, as they possess no special insight into moral matters and hence the control of technology policy through expert judgement represents an alternative form of technocratic control.

By asserting that we should adopt a bottom-up, citizen-led assessment of technology ethics we are presented with a challenge. We must find a way to facilitate deliberation on ethics in a manner which is both philosophically robust, in the sense of not simply being based upon knee-jerk reactions to moral problems, but also pluralistic, in that it incorporates a range of different perspectives, values and experiences. This chapter begins by discussing some potential solutions to these meta-ethical problems, and then ends with the presentation of a model of ethical deliberation grounded in the philosophy John Rawls's concept of "Reflective Equilibrium". In chapters 5, 6 and 7, this reflective equilibrium approach forms the basis of a methodology or decision-procedure for participatory-deliberative evaluation of technology ethics.

3.2 Resolving the Problem of Technocracy, Beginning with Habermas

The meta-ethical or *discourse ethics* of Karl–Otto Apel and Jürgen Habermas have been deeply influential in the theory and practice of deliberative democracy, and more recently, the design and implementation of participatory processes in civil society. Habermassian discourse ethics presents a theoretical effort to reformulate the insights of Kant's principles of deontology (concerning the moral obligations of the individual) in terms of the analysis of communicative structures. Kant believed that objective moral truth could only be deciphered within the rational cognitive processes of the moral agent, whereby, "...[everyone] must concede that the ground of obligation here must not be sought in the nature of man or in the circumstances in which he is placed, but sought *a priori* solely in the concepts of pure reason" (Kant 1785/1998). The Habermassian tradition asserts that the validity of a moral norm cannot be justified in the mind of an isolated individual reflecting on the world. Whereas Kant asserted that moral principles are extracted from the necessities forced upon a rational subject reflecting on the world, Habermas suggests that moral principles are extracted from the necessities forced upon individuals engaged in the discursive justification of validity claims, from the inescapable presuppositions of communication and argumentation (Habermas 1993). Discourse ethics concerns the externalising of what Kant termed the *dialogue interieur*, whereby the validity of a norm is justified not through the rational thought processes of the individual, but inter-subjectively in a process of argumentation between individuals as part of an interactive public deliberation or dialectic (Habermas 2002; Apel 1984; Habermas 1993), exchanging propositions and counter-propositions (between theses and antitheses) resulting in a synthesis of the opposing assertions.

The critical component of this Habermassian tradition is that of rational argumentation. Habermas asserts that moral actors are in possession of *communicative rationality*. Communicative rationality is the unconstrained, unifying, consensus-building force of argumentative speech; in which different participants overcome their 'subjective' views. In doing so, owing to the mutuality of rationally motivated conviction, they then assure themselves of both the unity of the objective world and the 'inter-subjectivity of their life-world' (Habermas 1984; Ajzner 1994). The individual's communicative rationality allows consensual moral action to be decided upon. Habermas believes that the roots of co-operation between moral actors in a deliberative process lie in the very structure of language itself. Built into language is the assumption that any speaker can evaluate, validate and defend his or her statements if needed. This ultimately amounts to an implicit commitment between one speaker and another to co-operate as without such rules the structure of language itself would be meaningless: agreements could never be met, jokes would not be funny and lies would be indistinguishable from truths. Perhaps paradoxically, if we did not assume that the utterances of someone speaking to us were true, then there would be no purpose in attempting to lie.

For Habermas, the language for political and moral decision-making occurs in the public sphere, "a discursive arena that is home to citizen debate, deliberation, agreement and action" (Villa 1992). In the public arena, such as that provided by deliberative decision-making processes, Habermas's meta-ethical position is founded upon creating an *ideal speech situation* founded upon a set of language rules. The following summary is derived from Habermas (1987):

1. Every subject with the competence to speak and act is allowed to take part in a discourse.
2. Everyone is allowed to question any assertion.
3. Everyone is allowed to introduce any assertion into the discourse.
4. Everyone is allowed to express his attitudes, desires and needs.
5. No speaker may be prevented by internal or external coercion from exercising his rights as laid down in 1, 2, 3 and 4.

3.3 Discourse Ethics and Participatory-Deliberative Decision-Making

Habermas's philosophy has been influential in a range of social sciences, and his theories have been consistently applied in the practice and analysis of deliberative decision-making. Webler in particular applies the theory of ideal speech to participatory approaches to environmental and technology decision-making, and introduces two supplementary concepts of *fairness* and *competence* in the evaluation of participatory-deliberative processes (Webler 1995). The concept of fairness implies that everyone should be provided with an equal opportunity to have a say in the process, decide upon its agenda, the rules of discourse and the discussion and also have equal and unrestricted access to knowledge and interpretations. Competence by contrast, is the so-called "meta-yardstick" of evaluating the discourse; it refers to the participants using all of the relevant information that is available at the time the decision is made (ibid).

In practical terms, the discourse ethics of this tradition presupposes that individuals can realistically remove political bias inherent to speech acts between moral agents. In deliberating upon technology choices and effects the Habermassian speech model assumes that the communicative rationality of the individuals (and the rules of their deliberation) within the ideal speech situation will allow consensual agreements to be made. Habermas's concept of rationality differentiates between *communicative* and *strategic* aspects. Communicative rationality is an understanding and acceptance of the better argument through co-operative use and understanding of language structures within a collaborative discourse. Strategic rationality, by contrast, is the ability to manipulate discourse through deploying strategies to influence the actions and understanding of other communicative actors. The distinction ultimately resides between action oriented toward mutual understanding (communicative rationality) and action oriented toward success (strategic rationality) (Johnson 1991).

Communicative rationality has been challenged by Habermas's opponents, notably Foucault. Foucault's (2002) critique of Habermassian notions of rationality is that a discourse can never be singularly defined as communicative, as it will always involve certain strategic elements; i.e. the content of a political or moral discourse is influenced by different actors deploying one or more strategic options or choices to their own advantage. Foucault insists that the very basis of Habermas's concept of language situations and communication is flawed and that the removal of strategic elements from (in this case ethical) discourse is impossible. The fact that this theory of communicative action is an idealistic deliberative theory is not, however, necessarily problematic in and of itself. Trying to develop rational discourse unhindered by the strategic manoeuvring of political actors is a laudable goal that reflects Habermas's commitment to the Enlightenment tradition; the striving of human nature for progressive improvement in moral character. The ideal speech situation is posited as a normative ideal, not a description of existing political practice, and these ideals provide a useful starting point for examining a deliberative process for ethical evaluation in Technology Assessment.

If we are to take the fostering of communicative rationality as one of the ultimate goals of a deliberative process involving citizen actors, then it is clear that as more strategic elements creep into the deliberative process then this will disempower them, as their influence wanes in the face of political power. At some point, therefore, public actors will logically cease to initiate change or additional communicative actions when they repeatedly lose out to strategic bargaining. As mentioned previously, if a governing organisation sets up a supposedly participatory-deliberative process that promises to facilitate communicative rationality and provide citizen and stakeholder actors with the opportunity to make decisions based solely upon the strength of rational argumentation, and then manipulates public discourse over technology to suit their own strategic ends, this may cause citizens to "feel that it is impossible to resolve political problems with the help of 'sincere' democratic debate" (Skolleerhorn 1998). In the context of the participatory-deliberative turn the notion of transparent, unbiased and open communication amongst stakeholder actors has become an intrinsic part of Technology Assessment. If however, as Foucault argues, it is impossible to truly achieve communicative rationality, how then can we realistically encourage open, fair and effective deliberation on ethical issues? Perhaps more fundamentally than that, however, is a meta-ethical question over the underlying assumption that actors possess a universal communicative rationality that is binary in nature. Communicative rationality is binary in the sense that individuals possess basic communicative rationality grounded in linguistic competence, implying that people are either rational or irrational, as if these were simple in/out descriptive categories. The second question then becomes, who can be considered rational and how is this decided upon?

3.4 Competing Rationalities

Within decision-making processes it is important to distinguish between different forms of rationality. In particular we must consider the difference between the *social rationality* of non-specialist citizen actors and the *bounded rationality* of experts (Perrow 1999). To return once more to the nuclear power example, engineers and risk managers planning a siting process for a new nuclear power station would likely adopt a somewhat utilitarian position basing their judgements upon available physical evidence, risk modelling and safety assessments to present a solution that is both rational and morally valid in that it reduces overall risks to the aggregate population in the immediate vicinity of the proposed site. However, local citizens affected by this siting process are likely to protest at such an imposition, and in turn, would highlight egalitarian and deontological normative principles, focussing upon the inequity of risk distribution between communities and the injustice of being forced to accept risks and other social and environmental burdens when other neighbouring communities are not. It is therefore 'rational' for them to criticise a policy that expects individuals to accept risks without clearly defined rewards. We are then presented with two rationalities, defined in one instance by an appeal to scientifically defined safety and the other upon procedural aspects of environmental justice. Deciding which form of 'rational argumentation' should win out between these two groups of deliberators is not easily resolved by appealing to the communicative rationality of participants, because the problem involves finding some way to choose between irreconcilable ethical principles.

Habermas sought to find solutions to such problems by generating consensual 'truth' from the communicative action of rational deliberative actors. The final goal of his ideal specch situation is *Verständigung* or 'shared understanding', as opposed to objective universal 'Truth' from meta-physical *a priori* moral rules. Rationality is the central pillar of this theory. A norm (ethical or otherwise) can only be accepted if *all* those affected can accept the associated consequences, to the extent that those consequences can be known (Habermas 1991; van Es 1998; Parking 1996). The questions is, whether Habermassian speech rules can alleviate deliberative conflict and allow competing sides to reach consensus. In practice within a deliberative decision-making process, we see Foucault's criticisms of Habermas played out, as competing rationalities will likely lead to entrenchment as each side seeks to convince the other of the superiority of the argument they propose. This is related to the aforementioned problem of negotiation and the inherent strategic aspects of communication - seeking a means with which to reduce political conflict and yet strengthen ethical legitimacy requires us to admit that rationality alone is insufficient to achieve a consensual outcome. I suggest that a potential solution to this problem may be to dispense with the notion that rationality is a pre-requisite for all forms of ethical evaluation; thus breaking from a paradigm that has long dominated Western moral philosophy. However, to do so requires significant meta-ethical justification.

3.4.1 Rationalism and Moral Emotions

There have been some serious challenges to rationalism in ethical deliberation. *Reason* has been the central tenet of moral philosophy since Plato. He presented a model of a divided self in which reason is firmly ensconced in the head where it rules over the passions, which rumble around in the chest and stomach (Plato 1949). Aristotle similarly conceived of reason as the wise master and emotion as the foolish slave whereby, "anger seems to listen to reason, but to hear wrong, like hasty servants, who run off before they have heard everything their master tells them, and fail to do what they were ordered, or like dogs, which bark as soon as there is a knock without waiting to see if the visitor is a friend" (Aristotle 350 B.C./2000). From the foundations of these early philosophical writings, Western philosophy has tended to focus upon moral reasoning, whereas the moral emotions have been regarded with a degree of suspicion (Solomon 1993; Haidt 2003).

Notable critics of rationalist ethics such as David Hume were convinced that moral judgements were always mediated by emotional considerations, and are therefore non-rational, though Hume did not ascribe any normative weight to the emotional reactions of moral agents. The attribution of normative weight to emotions has occurred more recently, with prominent figures in modern philosophy such as Leon Kass (former US presidential advisor on bioethics) who writes of the importance of disgust, repugnance or yuckiness that people feel towards certain actions or policies (particularly in regard to biotechnologies) as being implicit elements of a type of moral wisdom. To Kass some technologies such as stem cell research or synthetic biology, violate the moral dignity of agents, and thus their reactions of disgust are indicators of a means to make ethically valid decisions. Critics of Kass, notably Harris (2004) and Evans (2010) suggests that such thinking rests upon an ontological mistake, as it simply conflates Hume's Is/Ought distinction. Kass confuses what people believe to be right or wrong with an evaluation of what ought to be right or wrong based upon sound moral premises.

Though easy to dismiss such category errors in moral thinking, there a number of significant challenges to the idea that ethics must be implicitly rational if it is to be trusted. The first challenge I present to the accepted role of reason in ethics, comes not from philosophy, but from the cognitive sciences, social and moral psychology, and evolutionary biology. Normative ethics asserts that the reasoned individual performs (or should perform) moral decision-making via a conscious application of meta-physical principles. However, recent research in the cognitive science of moral reasoning suggests that the mental processes of moralising are in fact very different. Researchers have shown not only that much of human cognition (overall) occurs automatically and outside the scope of consciousness (Bargh and Chartrand 1999), but also that people are often not very adept at describing the process of how they actually reached a particular judgement (Nisbett and Wilson 1977).

3.4.2 Automaticity

An important psychological concept to our understanding of how people arise at ethical judgements, is that of "automaticity". Automaticity describes skilled actions that people develop through repeatedly practising the same activity – an obvious example being how individuals learn to drive a car. The repetition of physical actions result in the capacity to effortlessly complete everyday tasks with low interference of other simultaneous activities and without conscious thought to step-by-step process (Schneider 2003; Schneider and Chein 2003). Some skills can therefore appear to emerge subconsciously after a period of practice. The concept of automaticity has been applied by cognitive moral psychologists to describe the mind's ability to 'resolve' many moral problems, and produce moral judgements, unconsciously and automatically (Greene and Haidt 2002). Haidt (2001) suggests that instead of accepting a deliberative or dialectical model of moral cognition, we adopt a *social intuitionist* model of moral 'automaticity'. Social intuitionism stresses that ethical judgement is somewhat like aesthetic judgement; we see an action or hear a story and we have an instant feeling of approval or disapproval. Moral judgements on an individual level are conceptualised as affect-laden intuitions - they appear suddenly and effortlessly in consciousness with an affective 'valence'; i.e. certain actions, situations and beliefs feel 'right' or 'wrong', but the individual arrives at the judgement without any feeling of having gone through the steps of searching, weighing evidence or inferring a conclusion.

This theory of a socially intuitive moral psychology proposes an interesting challenge to the traditional rationalist theories coming out of the 'cognitive revolution' of the 1950's and 1960's. During this period, the dominant behaviourist and Freudian theories of the early 20th century gave way to 'mental models' and information processing as the preferred framework in psychology. Notably the works of Piaget (1965) and Kohlberg (1969) deal with how humans developed their cognitive reasoning about ethical issues and concluded that human moral psychology develops in a progressive fashion; in three principle stages of moral progression.

The first stage is Kohlberg's (1984) notion of a 'pre-conventional' level of moral thinking. This first stage, he argued, is that generally found in children at the primary school level. In pre-conventional moral psychology, individuals behave according to socially acceptable norms simply because they are instructed to do so by an authority figure such as parent, carer or teacher. Obedience to these moral norms is compelled by the threat or application of punishment if the individual transgresses. The morality of an action is judged in relation to its direct (and immediate) consequences. The concept of self is composed in an egocentric manner, as the individual has not yet adopted or internalised societal conventions regarding on right or wrong, but instead focuses largely on external consequences that certain actions may bring. Thus progression within this preconventional level is characterised by a view that 'right' behaviour involves acting in one's own best interests.

The second 'conventional' level of moral thinking is that generally found in the general 'society at large'. Individuals within the conventional level adopt an attitude which seeks to do what will gain the approval of others, generally either peers or superiors, and so the fulfilling of social roles and what it means to be perceived as 'good' or 'bad' dominates moral thinking. Progression within this second stage involves the individual orienting their moral thinking towards abiding by laws, rules and social conventions and thus responding to the obligations of duty that these entail. Most active members of society remain at this stage, whereby morality is still predominantly dictated by an outside force (Kohlberg 1973). The third level of 'post-conventional' moral thinking is one that the majority of adults never reach. The preliminary stage of post-conventional moral thinking is an understanding of social mutuality and a genuine interest in the welfare of others. The world is viewed in terms of value pluralism, an understanding that different people hold different opinions, rights and values and that each must be mutually respected as being held unique to the individual holding such values and the culture from which those values stem. Laws, rules and regulations are thus regarded as necessary social contracts rather than monolithic edicts. Those rules that are contrary to the welfare of society (or indeed for the welfare of minorities within the general populace) should be changed when necessary. Progression within this stage leads to an individual's respect for universal principle and the demands of individual conscience (Kohlberg 1984). This last stage is similar to the ideal of communicative rationality suggested by Habermas – at this stage the individual develops a truly 'philosophical' understanding of ethical principles, whereby logic and rational argumentation shape moral deliberation and understanding, rather than uncritical acceptance of whatever dominant discourse of ethics influences the individual at the time.

The social intuitionist model differs from Piaget and Kohlberg. Although it allows for higher cognition it nevertheless suggests that moral judgements are produced primarily by emotional and 'non-rational' processes rather than deliberative, dialectical and rational ones; a fact that Kohlberg's work overlooks. More significantly perhaps is that in the social intuitionist model, the process of moral reasoning is relegated to the role of making *posthoc* justifications for antecedent moral judgements (Pizarro and Bloom 2003). An individual's moral judgements emerge on an affective or emotional level and are then later justified within a framework of rational moral reasoning in order to provide external validity. The affective and emotional facets of moral cognition present a challenge for normative ethics. Social intuititionism appears to confirm the empiricist philosopher David Hume's argument that moral beliefs are ultimately psychological rather than logical or empirical, an expression of emotion; of "the passions". To Hume there is nothing logical, teleological, rational or divine about morality; it is so reducible to human feeling alone, that; "…'tis not contrary to reason to prefer the destruction of the whole world to the scratching of my finger" (Hume 1739). In respect to normative moral philosophy, however, it is important to reflect upon critiques of this position.

Held (1996) and Miller (2000) argue that normative ethics should not be subsumed into descriptive ethics by way of the assertion that morals are simply controlled purely by subconscious cognitive processes. To do so, implies that moral philosophy lacks critical value for encouraging individuals to arrive at morally reflective judgements, and would conflate the normative with the empirical. Moral values should not be defined solely as personal preferences, or conflated with other non-ethical cultural, religious and political values. Though the psychological research shows that we are not entirely rational moral actors when making decisions, this does not mean that we should accept an extreme relativist position that reduces all moral assertions to simple statements of personal taste or reflections of dominant cultural discourse. In some respects the naturalisation of ethics and the growing influence of moral psychology undermines the role of philosophical thinking. To adopt the social intuitionist position wholesale would diminish the critical and evaluative edge that moral philosophy provides, but it is important to understand that though these two ways of understanding morality remain distinct, they can be complementary. The purpose of the latter is not simply to describe morality but to facilitate critical reflection in order to *improve* the ethical validity of decision-making for the individual, thus improving the underlying quality of individual judgements that appear to emerge from the subconscious mind. I argue, therefore, that it is important to find a reflective balance between these two aspects, the descriptive and the normative (a problem lying on well-trodden ground from Hulme's Is-Ought conundrum onwards). In doing so, we can develop a PTA process that satisfies both the philosophical criteria for ethical acceptability and the political requirement for widespread engagement and public decision-making support.

3.4.3 Incorporating Rational and Non-rational Ethical Judgements

In searching for this balance, we have on the one hand, the assertion that moral judgements are simply the unconscious processing of our reactions to the world around us; that particular technological strategies are morally wrong because they *feel* wrong. Any ontological justification that a particular strategy is 'right' or 'wrong' is construed as being merely coincidental to the moral judgement itself. On the other hand we have the guiding normative principles of ethical theory that assert there are absolute 'rights and wrongs' on the basis of meta-physics and rational deliberation, argumentation and justification. The conceptual framework informing this book is based upon a search for balance between these two positions - between emotion and rationality, relativism and absolutism, and between descriptive and normative ethics. In finding the means to balance these aspects, I suggest a framework based upon the concept of reflective equilibrium.

3.5 Reflective Equilibrium and Its Critique

Reflective equilibrium originated in the work of Goodman. He proposed an approach to the 'justification by balance' of rules of inductive logic that involve justifying the rules of inference in inductive or deductive logic by bringing them into reflective equilibrium with what we judge to be acceptable inferences in a broad range of particular cases (Goodman 1955). The term was introduced and applied to moral philosophy, and then broadly popularised by John Rawls. He applied it as a complementary theory to the *Original Position* in his work 'A *Theory of Justice*' (Rawls 1999). Reflective equilibrium involves an individual working back and forth between considered judgements about specific instances or particular cases, the normative principles or moral rules that are believed to govern them and the theoretical considerations believed to bear on accepting these considered judgements, principles, or rules; revising any of these elements wherever necessary in order to achieve an acceptable coherence among them (Cohen 2004). The goal is to find coherence among judgements, principles and theoretical considerations. It is ultimately the end-point of a deliberative process in which an individual reflects upon and revises their beliefs about an area of moral inquiry. In practical terms, it involves the specification, reciprocal weighing, testing, revising, and balancing of principles, rules, background theories, and particular judgements. It must be noted, however, that this reflective equilibrium need not remain stable, as individuals undergoing the process may modify it as new elements arise in their thinking (Schroeter 2004).

Reflective equilibrium balances judgements that are 'bottom-up' (which in this case could be those judgements expressed by citizens or stakeholders) without critical or theoretical evaluation and principles that are theory driven, based in meta-physics and essentially 'top-down' in nature. It has been developed as a methodological instrument for ethical theory development, in order to obtain a *coherent* ethical theory that is sensitive to the 'facts of moral life'; standing in direct opposition to a top-down applied ethics approach which essentially tries to plug facts into principles (Daniels 1996a). Reflective equilibrium is by contrast a flat-structured ontological position. The relationship between principles, theories and judgements must be one that balances according to the relevance of principles to inform the case and the specificities of the case to amend the principles used. This is the reflective aspect of the equilibrium - one thinks about which judgement a principle might require of them and about which principle could accommodate a particular judgement or stance on a particular issue, and then cycles between the two, refining both iteratively.

The procedure involves considering variations on the particular case, testing the principle against them and then refining and specifying the principle to accommodate judgements made about these variations. Those deliberating might also revise their judgements about certain cases if the initial views do not fit with the principles they are inclined to accept. As Daniels (1996a) argues, such a revision may constitute a moral surprise or discovery, implying that it is a learning process as much as an analytical one. By synthesising new moral positions the procedure allows some creativity into the moral evaluations, rather than the conservative tendencies inherent to applying top down normative ethical theories.

In practice, individuals clarify their particular moral judgements about an issue by looking for the coherence of those judgements with their beliefs about similar cases and about broader moral and factual issues, thus they have sought reflective equilibrium as a way of clarifying for themselves what they ought to do (Daniels 2003). It is 'reflective' in the sense that one knows to what principles one's judgements conform, and 'equilibrium' in the sense that principles and judgements coincide. This process creates what is termed 'narrow' reflective equilibrium, one that coherently balances moral judgements and the theoretical underpinnings that support or contradict those judgements.

Though the model of reflective equilibrium shows great promise for the development of a decision-procedure that balances between citizen moral judgements and broader ethical principles in PTA, it has been subject to significant critique. Some, such as Hare (1973) and Brandt (1979), have argued that the considered moral judgements or intuitions that people bring to the reflective process lack initial credibility. These critics have questioned whether judgements which are not based upon a priori principles provide a sufficient epistemological basis or grounding on which we can seek justification within ethical decision-making. They suggest that an act of simply making a set of beliefs (that lack this initial credibility) into a coherent balance cannot produce justification, because the pre-theoretical intuitions (what I term bottom-up moral judgements) upon which they are based are simply a product of social, political and cultural indoctrination and so they reflect bias, superstition, or mere historical accident. Similarly, judgements lack evidential force regarding a moral order and so coherence in reflective equilibrium is dependent only upon a kind of persuasiveness, one that comes from coherence among many elements being more convincing than the conviction that comes from any of its parts (Brandt 1990). Brant, in essence argues for a process of formulating moral judgements that is based upon the interrogation of moral principles based upon 'facts and logic' rather than feelings, intuitions and fallible social values. Lyons (1989) takes a similar line of argument:

> "... The justificatory force of coherence arguments is unclear. Suppose one assumes that there are such things as valid principles of Justice which can be justified in some way; suppose one believes, moreover, that a coherence argument explicates our shared sense of justice, giving precise expression to our basic moral convictions: one may still doubt whether a coherence argument says anything about the validity of such principles."

These two criticisms are founded on an inherent ontological position that intuitions are fallible and thus cannot be considered as indicators of moral truth. In essence, by starting from a point of intuition, the whole process is founded on subjective beliefs and is thus unreliable. Defenders of Rawls, most notably Daniels (1979), have argued that reflective equilibrium's value lies not in its ability to justify intuitions as the foundations of moral truth (what might be termed 'pure intuitionsim'), rather its value lies in the variety of alternative viewpoints enlisted to encourage the examination and possible revision of initial judgements (see also Wood 2012). It is therefore a form of procedural or deliberative ethics that encourages personal reflection upon moral judgements in relation to established principles and has the capacity to re-contextualise and reconfigure principles in light of intuitions. A rather practical and common sense defence of Rawls's model is simply to state that no individual begins moral inquiry from a perspective outside of their established belief system, norms and pre-existing values. Ethical inquiry (like its scientific counterpart) is never value free and performed in a social and political vacuum. Prior moral judgements are always influential in the development of any ethical theory or the application of principles to cases, and removing these elements completely is impossible. In defending the coherentist approach one could simply state that it would be fruitless to build an ethical decision-model that pretends otherwise, and so the 'strong' epistemic critique of intuitionism falls down in relation to practical ethical decision-making contexts.

3.5.1 Wide Reflective Equilibrium

Critique of the coherentist reflective equilibrium model is further complicated by the distinction between *narrow* and *wide* reflective equilibrium. When we focus solely upon specific cases or issues and a group of selected principles that apply to them, and do not subject the views we encounter to extensive criticism from alternative moral perspectives, we are seeking only *narrow* reflective equilibrium (Rawls 1974). *Wide* reflective equilibrium by contrast, is the process of bringing to bear the broadest evidence and critical scrutiny we can, drawing on all the different moral and non-moral beliefs and theories that are arguably relevant to our selection of principles or adherence to our moral judgements (Daniels 1996a). It aims for maximal coherence or 'fit' between an individual's considered moral judgements, a set of moral principles and relevant background theories (including non-ethical ones). In defending reflective equilibrium, Daniels argues that this process provides a method for constructing or selecting the ethical theory that is authoritative and superior to its competitors because the process of broadening out reflective equilibrium from universal theories and moral judgements to real world situational ethics provides solid justification for accepting the coherentist approach.

Essentially *narrow* reflective equilibrium operates on a practical level, it is a process through which individuals can reflect upon particular issues or cases; whereas *wide* reflective equilibrium is essentially meta-ethical, it is the justification of which norms and principles can be used within the narrow reflective equilibrium. Reflective equilibrium in a more general sense, is an iterative and highly inductive form of reasoning used in building a coherent balance between moral judgements, theories and principles by considering a particular considered judgement in a particular situation; one tries to develop a more general rule and to link that both to other practical judgements and to a higher level background theory. Daniels's answer to reflective equilibrium's critics is to 'put it into action... and let it be judged by its results' (Daniels 1996b). This focus upon the practical value of reflective equilibrium is important, because we are asked to assess the model not only from the basis of a priori principles and norms, but from its use as a practical tool to make ethically informed decisions. With this in mind, I take the reflective equilibrium model and assess it through the lens of philosophical pragmatism: a philosophy concerned with the practical values of ideas in real-world decision-making.

3.6 Pragmatism, Reflective Equilibrium and Technology Assessment

All of the difficulties, antagonisms and dichotomies that have been presented so far, share one common underlying feature; they are all, in essence, ontological problems that stem from a set of fundamental dualisms. Philosophy is littered with such interrelated dualisms, going back to Descartes's distinctions between mind and body, fact and value, object and subject. They are thoroughly integrated into what we might term a Western understanding of the physical and social world, reflecting the way different theorists believe social reality and knowledge production is (or should) be created, evaluated and applied. In this last part of the chapter, I question whether these ontological dualisms are necessary in defining a robust technology ethics. I argue that it is appropriate and necessary to break apart such dualisms in order to reveal something new about ethical decision-making.

The underlying epistemological position within this book is that an ethics-centred PTA utilising a monistic ethical framework is fundamentally flawed, both conceptually and practically in terms of implementing politically legitimate and publicly acceptable technology decisions. I have thus far presented a model of applied ethics characterised as an application of 'top-down' theories to real world contexts; but it is important to note that this is by no means a universal feature of the applied ethics literature. Indeed, philosophers such as Alasdair MacIntyre and Tom Beauchamp have questioned whether this is a useful conceptualisation at all. They argue that it is a mistake to think of ethics as a body of theory that can be brought in, when necessary, to sort out any particularly 'real world' dilemma (MacIntyre 1984; Beauchamp 1984). Though to many philosophers the concept of applied ethics implies a separation of theory and practice - that theorising takes place first and is then put into practice - there are others that have sought to move towards a system of applying ethics in a manner that operates as more of a 'two-way street'. Such an ethics involves using theory to inform practice and crucially, to allow practice to inform theory: in essence replacing a dualism with a dialectic. To justify such an applied ethics I turn to the tenets of philosophical pragmatism for support.

3.6.1 Pragmatism

In its broadest and most familiar sense, the term 'pragmatism' refers to the usefulness, workability and practicality of ideas as being the central criteria of their merit. The term 'pragmatic' in common use has both positive and negative connotations. A term often used to describe people and their actions; a pragmatic person is one who is level-headed, down-to-earth, a doer rather than a thinker. To a normative ethicist this could arguably be a hindrance rather than a benefit. Focus upon the practicalities of ethical problems rather than reasoning, abstraction and logic is perhaps unusual in moral philosophy. Pragmatism in everyday parlance is often perceived as a beneficial quality. A pragmatic person is one that focuses upon 'what is' and 'what can be done' rather than (perhaps fruitlessly) reflecting on what 'should be'. If we were, however, to follow this definition to the extreme, we would advocate a type of pragmatism that is simply unreflective practice, or a type of anti-intellectualism. *Philosophical pragmatism* is neither of these things.

Pragmatism as a branch of philosophy is rather different to its commonly used definition. It could be considered a uniquely North American tradition in philosophy. The original pragmatists of the late 19th Century such as Charles Sanders Pierce, William James, George Herbert Mead and John Dewey had extensive influence on American and later international philosophy, influencing the works of Willard Quine, Donald Davidson, Hilary Putnam, Richard Rorty and Jürgen Habermas. It was not, however, well received in Europe as a whole, as it was broadly perceived as opportunistic and superficial, partly due to the tendency towards consequentialism and meliorism (holding a general belief that the world tends to become better over time and that humans can aid its betterment). Though initially unpopular in European philosophy, pragmatism has made something of a resurgence, particularly in the fields of environmental and technology ethics. This resurgence is due to an understanding of the complex and uncertain nature of new threats that the world faces. Issues such as climate change, ecosystem conservation, sustainable agriculture, risk bearing technology management and natural resource use require practical action informed by, but not substituted with philosophical deliberation (Light and Katz 1996).

In relation to the novel environmental and technological risk challenges of the late modern 'risk society' (Beck 1996) there has been a tendency among philosophers to turn to standard ethical theories and principles for guidance in solving new challenges – and as the need arose, to apply the theories to practical matters (Des Jardins 1997). Pragmatism by contrast is concerned with a search for new ethical theories and approaches; and has been particularly influential within debates over the ethical assessment of environmental and technological risks at a time when the traditional normative approaches of rights or utility have been frequently criticised for their anthropocentric bias (Wenz 2001; Sylvan 2003; Minteer 2001). Pragmatism by contrast, focuses on the meaning and value of an idea in relation to the practical consequences of its implementation (Rosenthal 1994); breaking down the dichotomies that pervade philosophical arguments. The divides between objectivity/subjectivity, fact/value, deontological/utilitarian ethics etc. are broken apart in order to find practical solutions to philosophical problems and thus avoid the trap of conservative normative ethical analysis principally concerned with which corpus of moral rules to choose and then apply. As William James suggests, "...there is no such thing as an ethical philosophy dogmatically made up in advance" (James 1979), the goal of pragmatist ethics, therefore, is to make normative ethical theory open to modification when the appearance of novel moral problems in practice demands it (Parker 1993).

For the pragmatist, ethical thinking takes place in the context of moral practice — in intelligent, shared, and imaginative engagement with actual situations, attentive to the details and to the new possibilities they may open up, rather than seeking a metaphysically justified 'final' analysis. Thus, pragmatism is primarily focussed upon the meta-ethical considerations surrounding the processes of moral inquiry rather than simply in the products of normative reasoning (Caspary 2000). A pragmatically justified course of action is discovered empirically, though this may be in the form of 'trial and error' rather than the formal experimental models of positivist scientific inquiry. This empiricist stance construes ethics as specific only to the particular situation, within particular temporal and spatial horizons of action. A pragmatic method of ethical reflection may influence decision-making by utilising a complex network of scientific, economic and normative judgements to generate practical solutions to moral problems. It does not assume that those solutions are generalisable to all situations, or that the judgements are fixed, abstract and immutable. To the pragmatist, moral decisions are by contrast specific, particular and open to reinterpretation and change. In deciphering these moral solutions to complex ethical issues, normative theories may indeed prove useful, but only as tools to be used to evaluate the situation, not as ends in themselves (Farber 1999).

Light and Katz's volume on environmental pragmatism is of particular note in this respect, as it focuses upon achieving what they term meta-theoretical pluralism aimed at opening environmental policy-making to the "plausibility of divergent ethical theories working together in a single moral enterprise" (Light and Katz 1996). In terms of practical application, Thompson (1996) and Varner et al. (1996), within the same volume, argue that pragmatists might endorse ethical decisions based upon rights or utility although the philosophical justification will be procedural, and hence not an endorsement either of rights-based or utility theory; as the application of ethics to practice is not a question of applying the correct theory to a specific situation. Within a pragmatist framework of evaluation, ethical theories can be used as tools to sharpen and clarify positions and clearly delineate the terms of the debate. Pragmatist ethics can therefore be summarised as a means to construct new possibilities for moral action through highlighting the creative character of finding solutions to moral problems (Joas 1993) rather than the application of pre-given normative rules to a 'real world' situation. Thus pragmatists argue that philosophy should be used as a force for practical political change to the way that individuals engage with their social, natural and technological environments and make decisions about how to proceed.

3.6.2 The Tenets of Philosophical Pragmatism

William James's work, 'Pragmatism: a new name for some old ways of thinking' (1907) presents a coherent outline of the defining features of a pragmatist philosophy. To summarise he states:

Pragmatism is a method of justification, not a theory with a fixed content.

- It is an empiricist tradition i.e. it lies predominantly on empirically given phenomena.
- Philosophical reasoning and scientific reasoning share a common structure – i.e. both represent a grounded search for useful generalisations and explanations.
- Pragmatism is non-reductionist i.e. takes into account a broad array of phenomena without reducing it down to one or two core notions.
- It attempts to do justice to the variety of human experience
- Pragmatic justification is coherentist, and consists in an ongoing process of integrating new assumptions into a larger body of knowledge.
- There is no fundamental difference between thinking and other human activities – whether it is the truth of thinking or the rightness of moral action at stake, in all cases it is the practical success of the activity that is its criterion of acceptability.

Pragmatism is essentially consequentialist. Analysis tends to focus upon the outcomes of ethical actions rather than the specific moral intentions of ethical actors. It is not, however, synonymous with the consequentialism of the utilitarian philosophers; the consequentialism of pragmatism is based upon action while utilitarianism emphasises usefulness as the primary criterion of ethical validity. Its consequentialism is meta-theoretical rather than normative. It is concerned with the context of putting theories into practice rather than generating a substantive set of new concepts for defining the rules of the social and moral world, hence it is a very broad church.

Pragmatism is a means of clarifying one's position through focus upon the end point of moral reasoning, thus it is broadly a method of moral reasoning rather than a doctrine, principle or corpus of rules. Despite the diversity of pragmatism, there are a number of shared underlying epistemological assumptions. In particular there is a focus upon the consequences of ideas to the practice of moral action; an assertion of the importance of an experimental attitude - testing the practical implications of applying philosophical tenets to real world cases; a fallibilist stance – accepting that our convictions are of a provisional nature and are in principle susceptible to repeal or review; and an anti-sceptical stance - understanding that the value of knowledge is based upon its practical application rather than its ontological validity. Pragmatism focuses upon grounding knowledge upon a series of postulates rather than universal 'truths', we must therefore rely upon a set of propositions that are accepted as true in order to provide a basis for logical reasoning.

A central tenet of a (general) pragmatist philosophy is therefore that something is true if it useful to believe; so there is an insistence upon practicality as a component of meaning and truth. According to James, to say that a belief is true, is to say that the belief succeeds in making sense of the world and is not contradicted by experience (James 1978). Pragmatism rejects the view that human concepts and intellect can solely and accurately represent reality, and so it stands in opposition to positivism and rationalism: asserting that only through the struggle of intelligent organisms with the surrounding environment can theories acquire significance and that only with a theory's success in this struggle does it become true.

Technology assessment is a prime example of how a pragmatist philosophy can benefit practice. One of the key 'concrete problems' in the assessment of SECT, is that of uncertainty in both its technical and social dimensions. To the philosophical pragmatist there are no fundamental moral truths that will remain unchanged; hence striving for an absolute and immutable ethical ideal is fruitless. We must get along without certainty, by solving practical, not theoretical problems and by adjusting the ends we pursue with the means available to accomplish them. Otherwise "method becomes an obstacle to morality, dogma the foe of deliberation, and the ideal society we aspire to in theory will become a formidable enemy of the good society we can in fact achieve" (Sagoff 1988). From a pragmatic perspective, we cannot make up our theories and rules in advance, they must be open to modification when we are faced with novel practical moral problems. The idea that technology produces novel moral problems is important. With an ever-changing and developing technological culture the issues of moral importance will continue to shift and interact. An absolutist, top-down applied ethics is fallible in this respect because abstract ethical maxims are unresponsive to socio-technological change. Pragmatic ethics seeks to provide this flexibility.

3.7 Conclusions

In the last two chapters I have presented a range of meta-ethical challenges to the notion that non-specialist citizen actors can evaluate the ethical issues inherent to the assessment of SECT. The epistemological position presented in this book prioritises the evaluation of ethical issues from the 'bottom-up'. We must find the means to elicit the moral judgements that citizens and stakeholders hold, assess them in light of a range of moral principles and then provide the means to balance between these different facets in a manner that is both coherent, contextually situated and practically relevant to technology policy and development. Rawls's model of reflective equilibrium is precisely this form of balancing approach. By starting from a position of outlining moral judgements, it draws upon an intuitionist model of ethical assessment, trusting in the practical rationality of individuals to imagine and envisage a course of action that is ethically legitimate. In guiding this process of ethical decision-making Rawls's model employs ethical principles in concert with judgements. The iterative process of comparing considered judgements in light of principles and reconfigured and situated principles in light of case-specific contexts and moral intuitions is a fruitful model of ethical decision-making that is, I would suggest, entirely compatible with a pragmatist epistemology. The emphasis upon practice however, requires us to develop this theory into a useable model. What we require are 'ethical tools' – procedures that encourage individuals with no background or training in ethics to critically reflect upon their judgements; judgements that are affect-laden, emotional positions. We must then create the means to allow them to reflect upon the validity of those positions and then relate them back to real-world problems, where they can develop and recommend potential practical solutions to socio-technical problems. Wide reflective equilibrium is the approach through which this outcome is sought because it requires a commitment by participants to utilising iterative and inductive reasoning and reflection upon the ethical aspects of the problem, whilst helping to frame their deliberations procedurally. Ethical justification within the proposed pragmatist framework emerges through coherent deliberation amongst participants about judgements and principles rather than the application of normative rules. Reasoning about ethical issues requires reflection upon the technical, political, legal and socio-economic contexts and policy practices currently in place; the affected stakeholders both human and non-human - currently alive and in the future; the theoretical frameworks and principles that govern our understanding of right and wrong action; the issues and their conceptualisation as being morally contentious; and the personal moral reactions and judgements of the participants. This presents a complex and challenging picture, not only from a philosophical perspective, but also a practical and methodological one.

In the next chapter, I turn from the meta-ethical considerations of an ethical PTA procedure, towards more practical matters – namely, the techniques needed to simplify, clarify and organise these different, and often-times opposing, aspects. As Kaiser et al (2007) and Beekman and Brom (2007) suggest, what we require in such complex situations is a 'toolbox' of practical ethical procedures or techniques that make ethical judgements and the subsequent advice they give amenable to quality assurance and deliberative democratic transparency. Such a toolbox in this context would take the form of a series of participatory-deliberative methods designed to elicit, analyse and contextualise moral principles, judgements, theories and the issues to which they relate, presented in the framework of a structured discussion amongst citizen actors. In doing so, the following chapter assesses this emerging field of 'ethical tools' that have arisen primarily in the fields of bioethics and healthcare ethics. I then reflect upon their practical application to achieve the meta-ethical goals I have outlined here and in the previous chapter. Following this, I then synthesise a new methodological approach to ethical assessment based upon the strengths and limitations of these existing tools and those of more conventional participatory-deliberative methods.

References

Ajzner, J.: Some of the Problems of Rationality, Understanding, and Universalistic Ethics. Philosophy of the Social Sciences 24(4), 466–485 (1994)

Apel, K.O.: Understanding and explanation: a transcendental-pragmatic perspective. MIT Press, Cambridge (1984) (trans. Warnke, G.)

Aristotle: 350 B.C./2000. Nichomachean Ethics. Cambridge University Press, Cambridge (2000) (trans. Crisp, R.)

Bargh, J.A., Chartrand, T.L.: The unbearable automaticity of being. American Psychologist 54(7), 462–479 (1999)

Beauchamp, T.L.: On Eliminating the Distinction Between Applied Ethics and Ethical Theory. The Monist. 67, 514–531 (1984)

Beck, U.: Risk Society: Toward a New Modernity. Sage, London (1996)

Beekman, V., Brom, F.W.A.: Ethical Tools to Support Systematic Public Deliberations About the Ethical Aspects of Agricultural Biotechnologies. Journal of Agricultural and Environmental Ethics 20(1), 3–12 (2007)

Brandt, R.B.: A Theory of the Good and the Right. Oxford University Press, Oxford (1979)

Brandt, R.B.: The Science of Man and Wide Reflective Equilibrium. Ethics 100, 259–278 (1990)

Caspary, W.R.: Dewey on Democracy. Cornell University Press, Ithaca (2000)

Cohen, S.: The Nature of Moral Reasoning: The Framework and Activities of Ethical Deliberation, Argument, and Decision-Making. Oxford University Press, Melbourne (2004)

Daniels, N.: Wide Reflective Equilibrium and Theory Acceptance in Ethics. Journal of Philosophy 76(5), 256–282 (1979)

Daniels, N.: From Chance to Choice: Genetics and Justice. In: Summer, L.W., Boyle, J. (eds.) Philosophical Perspective in Bioethics. Toronto University Press, Toronto (1996a)

Daniels, N.: Justice and justification: Reflective equilibrium in theory and practice. Cambridge University Press, Cambridge (1996b)

Daniels, N.: Reflective Equilibrium. Stanford Encyclopedia of Philosophy (2003), http://www.seop.leeds.ac.uk/archives/spr2004/entries/reflective-equilibrium/ (accessed June 1, 2012)

Des Jardins, J.R.: Ethics, Energy, and Responsibilities to Future Generations. In: Des Jardins, J.R. (ed.) Environmental Ethics: An Introduction to Environmental Philosophy. Wadsworth Publishing Company, Belmont (1997)

Evans, D.: Emotions as Aids and Obstacles in Thinking About Risky Technologies. In: Roeser, S. (ed.) Emotions and Risky Technologies, pp. 81–88. Springer, Dordrecht (2010)

Farber, D.A.: Eco-pragmatism: Making Sensible Environmental Decisions in an Uncertain World. University of Chicago Press, London (1999)

Foucault, M.: The order of things: an archaeology of the human sciences. Routledge, London (2002)

Goodman, N.: Fact, Fiction, and Forecast. Harvard University Press, Cambridge (1955)

Greene, J., Haidt, J.: How (and where) does moral judgement work? Trends in Cognitive Sciences 16(12), 517–523 (2002)

Habermas, J.: Theory of Communicative Action. Reason and the Rationalization of Society, vol. 1. Beacon Press (1984) (trans. McCarthy, T.)

Habermas, J.: The Philosophical Discourse of Modernity. MIT Press, Cambridge (1987) (trans. Lawrence, F.G.)

Habermas, J.: Moral Consciousness and Communicative Action. MIT Press, Cambridge (1991) (trans. Lenhart, C., Weber Nicholson, S.)

Habermas, J.: Justification and application: remarks on discourse ethics. Polity Press, Cambridge (1993) (trans. Cronin, C.)

Habermas, J.: On the pragmatics of communication. Cambridge University Press, Cambridge (2002)

Haidt, J.: The emotional dog and its rational tail: a social intuitionist approach to moral judgment. Psychological Review 108, 814–834 (2001)

Haidt, J.: The moral emotions. In: Davidson, R.J., Scherer, K.R., Goldsmith, H.H. (eds.) Handbook of Affective Sciences, pp. 852–870. Oxford University Press, Oxford (2003)

Hare, R.M.: Rawls' Theory of Justice. Philosophical Quarterly 23, 144–155 (1973)

Harris, J.: Immortal Ethics. Annals of the New York Academy of Sciences 1019, 527–534 (2004)

Held, D.: Whose Agenda? Ethics versus Cognitive Science. In: May, L., Friedman, M., Clark, A. (eds.) Mind and Morals: Essays on Cognitive Science and Ethics. MIT Press, London (1996)

Hume, D.: Treatise on Human Nature: Of Virtue and Vice in General. Oxford University Press (1739)

James, W.: The Meaning of Truth. In: Buckhardt, F.H. (ed.) Essays in Philosophy: The Works of William James. Harvard University Press, Cambridge (1978)

James, W.: The Moral Philosopher and the Moral Life. In: The Will to Believe and Other Essays in Popular Philosophy. Harvard University Press, Cambridge (1979)

Joas, H.: Pragmatism and Social Theory. University of Chicago Press, Chicago (1993)

Johnson, J.: Habermas on Strategic and Communicative Action. Political Theory 19(2), 181–203 (1991)

Kaiser, M., Millar, K., Forsberg, E.M., Thorstensen, E., Tomkins, S.: Developing the ethical matrix as a decision support framework: GM fish as a case study. Journal of Agricultural and Environmental Ethics 20(1), 53–63 (2007)

Kant, I.: Groundwork of the Metaphysics of Morals (Grundlegung zur Metaphysik der Sitten). Cambridge Texts in the History of Philosophy. Cambridge University Press, Cambridge (1785/1998)

Kohlberg, L.: Stage and sequence: the cognitive-developmental approach to socialization. In: Goslin, D.A. (ed.) Handbook of Socialization Theory and Research, pp. 347–480. Rand McNally, London (1969)

Kohlberg, L.: The psychology of moral development: the nature and validity of moral stages. Harper & Row, London (1984)

Kohlberg, L.: The Claim to Moral Adequacy of a Highest Stage of Moral Judgment. The Journal of Philosophy 70(18), 630–646 (1973), doi:10.2307/2025030

Light, A., Katz, E.: Environmental Pragmatism. Routledge, London (1996)

Lyons, D.: The nature and soundness of contract and coherence arguments. In: Reading Rawls, Critical Studies of "A Theory of Justice", pp. 141–169. Stanford University Press (1989)

MacIntyre, A.: Does Applied Ethics Rest on a Mistake? The Monist 67, 489–513 (1984)

Miller, R.B.: Without Intuitions. Metaphilosophy 31(3), 231–250 (2000)

Minteer, B.A.: Intrinsic Value for Pragmatists? Environmental Ethics 23(1), 57–75 (2001)

Nisbett, R., Wilson, T.: Telling more than we can know: Verbal reports on mental processes. Psychological Review 84, 231–259 (1977)

Parker, K.A.: Public Hearings / Hearing Publics: A Pragmatic Approach to Applying Ethics. In: Society for the Advancement of American Philosophy Annual Meeting, The Pragmatism Cybrary, Vanderbuilt University (1993)

Parking, A.: On the Practical Relevance of Habermas's Theory Communicative Action. Social Theory and Practice 22(3), 417–442 (1996)

Perrow, C.: Normal Accidents: Living with High Risk Technologies. Princeton University Press, Princeton (1999)

Piaget, J.: The Moral Development of the Child. Free Press, London (1965)

Pizarro, D.A., Bloom, P.: The Intelligence of the Moral Intuitions: Comment on Haidt. Psychological Review 110(1), 193–196 (2003)

Plato: Timaeus. Bobbs-Merrili, Indianapolis (1949) (trans. Jowett, B.)

Rawls, J.: The Independence of Moral Theory. In: Proceedings and Addresses of the American Philosophical Association (1974)

Rawls, J.: A Theory of Justice, 2nd Aufl. Oxford University Press, Oxford (1999)

Rosenthal, S.B.: Charles Pierce's Pragmatic Pluralism. State University of New York, New York (1994)

Sagoff, M.: The Economy of the Earth. Cambridge University Press, Cambridge (1988)

Schneider, W.: Automaticity in complex cognition (2003), http://coglab.psy.cmu.edu/index_main.html (accessed June 1, 2012)

Schneider, W., Chein, J.M.: Controlled & automatic processing: behavior, theory, and biological mechanisms. Cognitive Science (27), 525–559 (2003)

Schroeter, F.: Reflective Equilibrium and Anti-theory. Nous 38(1), 110–134 (2004)

Skolleerhorn, E.: Habermas and Nature: The Theory of Communicative Action for Studying Environmental Policy. Journal of Environmental Planning and Management 41(5), 155–174 (1998)

Solomon, R.C.: The philosophy of emotions. In: Lewis, M., Haviland, J. (eds.) Handbook of Emotions. Guildford Press, London (1993)

Sylvan, R.: Is there a need for a new environmental ethic? In: Light, A., Royston III, H. (eds.) Environmental Ethics: An Anthology. Blackwell, Malden (2003)

Thompson, P.B.: Pragmatism and Policy: The Case of Water. In: Light, A., Katz, E. (eds.) Environmental Pragmatism. Routledge, London (1996)

van Es, R.: Negotiating Ethics as a Two Level Debate. In: Twentieth World Congress of Philosophy, Boston, MA (1998)

Varner, G.E., Gilbertz, S.J., Rai Paterson, T.: Teaching Environmental Ethics as a Method of Conflict Management. In: Light, A., Katz, E. (eds.) Environmental Pragmatism. Routledge, London (1996)

Villa, D.R.: Postmodernism and the Public Sphere. American Political Science Review 86(3), 712–721 (1992)

Webler, T.: "Right" Discourse in Citizen Participation: An Evaluative Yardstick. In: Renn, O., Webler, T., Wiedemann, P. (eds.) Fairness and Competence in Citizen Participation. Kluwer, Dordrecht (1995)

Wenz, P.: Environmental Ethics Today. Oxford University Press, Oxford (2001)

Wood, J.: The Global anti-fracking movement: What it wants, how it operates and what next. Control Risks Group, London (2012)

Chapter 4
Ethical Tools

4.1 Introduction

To encourage public and stakeholder deliberation on the ethical issues involved in technology implementation a number of novel participatory-deliberative tools, procedures and emergent frameworks have arisen to facilitate ethical deliberation amongst different actors with legitimate stakes in technology governance outcomes; though few have been adequately developed and tested to determine their applicability as public policy decision-support tools (Beekman and Brom 2007). This chapter presents three popular tools that have emerged in the fields of bioethics and healthcare ethics, each of which aims to fulfil such a purpose. After examining the philosophical grounding and policy applicability of the current raft of ethical tools, the remaining portion of the book then showcases a series of new deliberative decision-support tools that build upon their strengths and limitations in light of the foregoing discussion on philosophical pragmatism and John Rawls's concept of reflective equilibrium.

4.2 Ethical Tools

Kaiser et al (2004) suggest that any given procedure for analysing ethical issues in assisting policy-making must operate as a structured decision-support framework. It is through the application of these methods in practical decision-support or policy-making that they become practical ethical 'tools' (Beekman and Brom 2007). In relation to this practical applicability, Kaiser et al (2004) then go on to propose a set of criteria which such ethical tools must meet; namely they must provide:

- Ample substantive ethical content
- Good opportunities to facilitate transparent decision-making processes
- A multiplicity of stakeholder viewpoints, ethically relevant information and ethical arguments

M. Cotton, *Ethics and Technology Assessment: A Participatory Approach*,
Studies in Applied Philosophy, Epistemology and Rational Ethics 13,
DOI: 10.1007/978-3-642-45088-4_4, © Springer-Verlag Berlin Heidelberg 2014

In this chapter I present three existing ethical tools that fulfil these criteria, each emerging from the applied fields of bioethics and healthcare ethics. I then go on to evaluate their usefulness to the practice of PTA:

1. The Ethical Matrix
2. The Ethical Grid
3. The Ethical Delphi Method

4.3   The Ethical Matrix

The Ethical Matrix (hereafter EM) is designed specifically for the examination and assessment of ethical criteria in a given situation, such as a technological development, organisation or policy. Its creator Benjamin Mepham intended the EM to be a means of assisting people in making ethical decisions, particularly those that surround and permeate the introduction of new technologies into society. The EM was originally designed for the purpose of assessing agricultural production systems, such as the technologies and practices of dairy farming, from the perspectives of different groups affected by its employment (Mepham 1999; Mepham 1996), both as a teaching tool for students of agricultural ethics, and then later as a decision-support tool for policy-making and technology assessment.

The underlying rationale is that science and ethics are interconnected. Mepham (2003) argues that ethics is primarily a science of "how we should live"; consequently all technical and scientific issues impact upon this. Mepham's tool therefore appears promising for the analysis of ethics in a PTA context. He asserts that there are two ingredients necessary for the evaluation of the ethical impacts of technologies. The first is a set of prima facie principles and the second a list of agents 'that have interests', emphasising that ethical analysis requires a compromise between competing requirements. Analysis therefore needs to be (2005):

- Based in established ethical theory to give it authenticity
- Be sufficiently comprehensive to capture the main ethical concerns
- Employ user friendly language as far as possible

Mepham establishes the EM in normative theory by adopting Beauchamp and Childress's 'principlist' approach. Principlism is an extension of the Rawls's 'common sense rule' (Rawls 1951), applying four (in this case) prima facie ethical principles, which have been broadly accepted within their original field of medical ethics (Beauchamp and Childress 2001):

- Autonomy – respecting the decision-making capacities of autonomous persons
- Non-maleficence – avoiding the causation of harm
- Beneficence – a group of norms for providing net benefits
- Justice – distributing benefits, risks and costs fairly

What characterises 'common sense principlism' is its derivation not from specific normative ethical theories, but from a selection of principles that are commonly understood within society and thus have a broad degree of support from both ethical theories and cultural beliefs. The matrix then applies these principles to the deliberative consideration of specific practical questions involving a range of different stakeholder positions.

The supposed strength of principlism lies in the allowance of a stronger case based on one principle to outweigh a weaker case based on another in particular circumstances. This presents an alternative to monistic normative ethical theory approaches that tend to assert a single principle (or set of related principles) over others. Mepham applied specific principles according to the field of analysis (i.e. dairy farming) and chose stakeholders affected by the decisions in that sector. Recent revisions of the EM allow, however, for the substitution of different ethical principles to different cases. Applying the matrix to alternative fields changes the moral context, and consequently both principles and stakeholders can be amended based upon their relevance to the case.

The 'standard' EM substitutes the four Beauchamp and Childress principles for three, conflating beneficence and non-maleficence into 'wellbeing' - for simplification and because of the inter-relationship between preventing harm and enhancing quality of life. 'Autonomy' is kept, as is 'justice', although this was later re-labelled as 'fairness', in reference to the Rawlsian concept of 'justice as fairness' (Rawls 1999). The three principles are intended to represent three dominant philosophical perspectives in modern normative ethics: Kantian deontology, Benthamite utilitarianism, and Rawlsian social contract theory (Mepham 2005). He argues that principlism doesn't constitute an ethical theory in the strictest sense, nor does it use ethical theories, but is in fact a set of moral premises intended to clarify and assist deliberation. The EM avoids 'expert ethicist' reasoning by placing evaluation in the hands of 'non-experts'. Indeed the matrix was originally designed as a teaching tool, so simplicity, clarity and user-friendliness are its primary aims. Such simplicity is achieved by replacing (what is likely unfamiliar) philosophical terminology with commonly understood principles, while their grounding in established theory provides the basis for a robust analysis. In practice, the EM creates a grid format that shows the interactions between the principles and stakeholders. An example of such a matrix, showing the ethical issues of new build nuclear power is shown in Table 4.1.

For each cell of the matrix, the principle along the x-axis is applied to the interests of the 'stakeholder' along the y-axis, and the result is used as the basis for discussion. Thus, a plurality of perspectives is shown to some extent within the EM. There are at least four identified 'stakeholders' (by broadly conceptualising the term to include abstract elements such as 'the biosphere' or 'future generations'), so the needs and values of multiple groups can be represented. Similarly, the three ethical principles allow for some breadth of ethical debate and the production of an easily understandable tool for use by ethical non-experts. Ethical matrices have been used in a variety of contexts with different identified stakeholder groups and principles and hence with different inputs along each axis. Examples of empirical studies using ethical matrices in the academic literature include:

Table 4.1 Ethical matrix for new nuclear power

	Wellbeing	Autonomy	Fairness
Nuclear industry	Profit generation, growing employment	Freedom from regulation and planning constraints	Low cost electricity to consumers, alleviating fuel poverty
Citizens	Protection from risk of radiation leaks and accidents	Decision-making input to site selection	Compensation in the face of elevated risks
Future generations	An environment free of radiological contamination	Knowledge about past practices and impacts	Reciprocity across time frames, avoiding discounting of future lives
The biosphere	Environmental remediation of contaminated sites	Maintenance of biodiversity and ecological health	Non-anthropocentric valuation of natural resources

- Food production and commerce (Mepham 2000; FEC 2005)
- Novel or functional foods (which supposedly act like pharmaceutical products claiming specific health benefits to the consumer) (Chadwick et al. 2003; Mepham 1999, 2001)
- Fisheries management and genetically modified fish (Kaiser and Forsberg 2001; Kaiser et al. 2007)
- Forest management (Gamborg 2002)
- Animal farming and husbandry (Mepham 2003; Whiting 2004)
- Carbon capture and storage technologies (P. Boucher and C. Gough 2012)
- Transgenic animal farming (Small and Fisher 2005)
- Xenotransplantation (implantation of non-human organs into human hosts) (Kaiser 2004)
- Environmental remediation, restoration of radioactively contaminated areas and long-term management of radioactive wastes (D. Oughton et al. 2003; D.H. Oughton et al. 2003; Howard et al. 2002; Forsberg and Kaiser 2002; Cotton 2009).

4.3.1 Practical and Meta-Ethical Considerations

When applied to decision-making contexts for technology assessment and other forms of policy-making, the EM is intended as a tool for mapping out the issues underpinning a decision, rather than determining an ethical decision using some supposed metric of evaluation. By refraining from rule-making or adhering to ethical doctrine Mepham (2000) argues it is ethically neutral in its intent. Such neutrality is a requirement for pluralistic deliberation on ethical norms, moral values and their application to PTA. The EM therefore alludes to Habermassian discourse or procedural ethics mentioned in previous chapters, whereby the argumentation of moral principles by (communicatively rational) individuals ascribes ethical value to a decision. By considering a range of normative principles, the matrix seeks to remove philosophical bias in influencing the decision outcome. Oughton et al (2004) assert that the matrix helps to avoid bias towards specific moral values and addresses conflicts between different principles in a systematic way. However, even with all relevant information and systematic representation of different values, they recognised that moral judgement must be exercised, whilst also questioning who this moral judge should be. This is important because questioning the legitimacy of non-elected citizen representatives to act as 'moral judges' is itself an issue that requires meta-ethical justification.

In practice, the EM has been used in different ways by different implementing organisations. In some cases, such as Boucher and Gough's study of the ethics of carbon capture and storage technologies, the ethical matrix is used as a framing device for considering different ethical positions in a desk-based study of stakeholder perspectives on the technology, using a data-led process to construct a map of the ethical landscape i.e. emergent interpretations of various actors' ethical framings of the technology mapped across a range of moral principles (Boucher and Gough 2012). Though not strictly speaking a participatory-deliberative application of the method, it presents a relatively bottom-up model of the EM, in that it is led by different stakeholder ethical positions expressed in documents available in the public domain. In other studies a more active deliberative approach is taken. Gamborg (2002) suggests using the matrix in an expert-led consultation process involving a panel of scientific experts, members of local government, administrative agencies, private industry and members of the public. During consultation, a spokesperson from each group would "present their 'client's cases' (so to speak), in doing so outlining the pros and cons for each group". Each panel member and each member of the 'lay' audience is given a copy of the matrix. After the presentation of the case and ensuing discussion, participants indicate in each cell of the matrix, whether they feel that the ethical principle is likely to be upheld, violated or unaffected by the proposal. By collating these responses it is possible to obtain a verdict (ibid), i.e. a measurement of the prevailing ethical perspectives among the participants. In some respects this scenario is pluralistic, in the sense that it incorporates lay public responses in the matrix. However in this model, public-controlled ethical deliberation does not occur - only lay participant voting or weighting of a top-down matrix.

This proposal also highlights additional problems for participatory-deliberative decision-making, namely that many of the potentially affected stakeholders lack a mechanism for representation as many of the groups have no physical form and cannot take part in decision-making. Although not specifically a criticism of the matrix; many of the key affected groups identified in matrices used by different researchers such as 'animals', 'future generations' and 'the environment' are not stakeholder actors in the sense that they have no specific voice of their own. Others that could be included like 'the general public' or 'NGOs' do have a political voice, but their interests may be so diverse that they cannot be adequately represented by an individual spokesperson. Also, although it is plausible that some categories such as 'the Environment' can be represented by specific advocacy organisations, NGOs or interest groups (Greenpeace or Friends of the Earth for example); a meta-ethical issue remains around the extent to which proxy representatives can speak on the behalf of others, especially those that lack physical presence.

Though these different interests are difficult to represent in the EM, Schroeder and Palmer (2003) assert such categories of stakeholders like future generations and the environment must be included by default because these groups cannot intervene in the decision-making process and yet are deeply affected by the outcome. It is therefore necessary to identify and interpret the best means for assessing their needs and always include these 'groups' in deliberative decision-making. This raises the problem of a trade-off between meta-ethical validity and practical simplicity in an ethical tool such as this. For example, with each additional stakeholder group that is identified a new row is added, until it becomes too large and unwieldy for use as group discussion tool. Key stakeholders are therefore reduced down to universal groups such as 'local community'. This is problematic, however. Treating diverse groups as homogeneous entities (alongside others such as 'the general public', 'future generations' or 'the Environment'), firstly assumes that a potentially diverse group of matrix-using participants will all understand these monolithic categories to mean the same thing, and secondly, fails to express the diversity of values and interests within these labelled groups. By representing the stakeholder groups as isolated and homogenous categories, this may cause participants to bracket off the effect of group interaction. The problem being, that stakeholder groups tend to operate in a synergistic manner (O'Mahony, 2004); i.e. the ethical 'effect' of one group's actions strongly influences and affects the consequences for and behaviour of other related stakeholder groups. Although some principles (particularly Justice or Fairness) allow for discussion of the relationships within and between different actors, the matrix's design lacks a mechanism to illustrate and record such inter-relationships - it only records the relationship between a technology and each separate stakeholder in isolation.

A new design of matrix showing the intricate latticework of relationships between affected groups would increase the complexity of the model and again may lose the element of transparent simplicity. However, the notion of breaking out of the confines of a 3x4 (or 3x5) matrix is worthy of consideration. The development of tools for ethical assessment in analytic-deliberative contexts may therefore benefit from being based around more detailed mapping of the synergistic relationships between ethical values both within and among stakeholder groups - showing the interactive elements of stakeholder relationships and how these shape moral judgements.

4.3.2   The Choice of Principles

Similarly questions have been raised over the choices of the principles used and justifying the choice of any three principles over others. Again, the answer is grounded in part by the practical simplicity of the matrix. Having too many ethical principles makes the matrix cumbersome to use. If we were to justify three specific principles for the any specific case, we must question how to choose those which will provide the most informative exploration of the issues. Transposition from agricultural practice to other forms of Technology Assessment requires a re-evaluation of the ethical premises from which the analysis can take place. In some cases where the matrix has been used in decision-making, users have selected different principles. Alternatives such as 'dignity', 'rights', 'equality', 'fairness' and 'solidarity' etc. have all been utilised (Schroeder and Palmer 2003). However, if this process of principle selection is driven by experts or the facilitators who run participatory-deliberative processes then this raises a meta-ethical problem due to a 'framing effect', whereby ethical principle selection is predetermined by experts and hence 'top-down', in the sense I have used before. As I have argued, in PTA this is untenable. The function of a participatory decision-making process is to lead the analysis from the bottom up, i.e. from those (potentially) affected by the implementation of the technology.

In reference to this problem, Kaiser et al. (2007) developed a testing framework to compare a top-down ethical matrix (with facilitator or specialist defined principles) against a bottom-up (participant negotiated principles) matrix with lay participants. In the top-down workshop nine experts applied the matrix to discuss key issues raised by the development of GM fish. Broadly speaking, the experts concluded that the main problems with matrix were based upon the time constraints for discussion, the limitations of the knowledge of the participants and the requirement for a broader range of stakeholders to be involved in discussion particularly those with 'complementary backgrounds'. In written feedback however, "all participants believed the use of the Ethical Matrix helped the process" (Kaiser et al. 2007). The researchers also concluded that the workshop findings reinforced the perception that expert groups prefer to work with a top-down approach to implementing the EM. In contrast, their bottom-up approach involved less explicit facilitator guidance; deferring where appropriate to the majority views of the (usually) lay participants in specifying the principles and conducting ethical deliberation. The matrix was initially applied with the standard four principles (with 'Well-being' specified separately as 'Increased Benefits' and Reduced Harm, alongside 'Autonomy', and 'Fairness'). Participants then translated these principles into specifications for the interest groups and, following group discussion, 'Autonomy' was modified and 'Dignity' was subsequently used in the matrix. The participants also added additional stakeholders to the original list. Some argued for the inclusion of 'future generations' as a stakeholder group, although it was agreed that these considerations could be included under a 'Consumer' group. Others perceived 'Research and Knowledge Production' to be an important issue. As a result of this discussion, an additional stakeholder group, 'Research Community', was added to the matrix making a total of five (all from Kaiser et al. 2007).

The framing of the ethical debate through the predefined choice of principles by specialists or expert ethicists and their subsequent deliberation in the top-down (classical) EM is controversial for participatory decision-making processes, as this could potentially lead to criticisms of techno-centrism; albeit due to ethical rather than techno-scientific framing of the decision-problem. The bottom-up EM would therefore be preferable, although considerable ambiguity remains around how the principles themselves are chosen and how one set of principles is preferred to others. The justification of the choice of principles is an important meta-ethical concern. Unfortunately, the matrix lacks a specific mechanism for justification of principle selection and thus another tool is required for this purpose.

The EM's standard set of ethical principles are grounded in the dominant 'Western' themes of moral philosophy, originally designed to maximise the breadth of ethical debate. However, the terminology used to categorise these philosophical traditions as principles is itself open to question and the difficulty in translating this into meaningful deliberative discourse lies in the interpretation of the principles themselves. For example, 'Autonomy' could conceivably refer to rights, duties, self-determination, liberty, freedom from coercion and personal responsibility. It could also refer to the decision-making capacities of individuals, or the relationship between intentional agents and the constraints of societal institutions. Similarly, wellbeing can be interpreted on a variety of different levels, from the individual, communitarian, societal, or state levels. Justice could refer to legal processes of compensation, legal rights or political enforcement as well as Rawlsian, Hobbesian, Socratic or Aristotelian philosophical traditions. Although the matrix could be used as the means to elicit such discussions, it still lacks a mechanism for visually (and conceptually) clarifying different meanings – potentially causing confusion for both matrix users, and third parties evaluators of matrix-centred discussions.

One solution may be to stipulate precise principle definitions. Without this, the interpretation of each word as representing a broader theoretical category creates internal inconsistencies and potential conflict among stakeholder-participants using the EM, rendering a 'one-size fits all' ethical issue per stakeholder/ principle a rather limited analysis. The ethical impacts of different stakeholder groups are matched up to a single universal issue, so much information and ethical tension is lost (at the very least in the recording process) in the name of simplicity and keeping the matrix small enough to be a practicable tool.

4.3.3 Conclusions to the Ethical Matrix

The EM has been used to address the challenge of ethical deliberation in a variety of technological decision-making contexts. Despite its popularity however, some significant problems remain for its implementation as a deliberative decision-support tool in a process of PTA. The first significant critique stems from the inherent constraint of the 3x4 (or 4x4) design. This feature aids simplification and structuring of ethical discussions but also limits opportunities for creative problem solving outside of the matrix's pre-defined principle and stakeholder categories. The trade-off between free-flowing discussion and idea generation and structure and transparency is a persistent challenge for deliberative and inclusionary processes. To borrow Stirling's (2004) terminology, the creative problem-solving and idea generation aspects of participatory-deliberative methods (opening up) requires reining in at some point in order to 'close down' deliberation and reach conclusions.

Mepham (2005) argues that basically, the matrix represents a checklist of concerns structured around ethical theory, and at best, allows for the stimulation of structured ethical debate from a range of perspectives. To open up decision-making, effective bottom-up deliberation is necessary: participant control and ownership of the process mitigates the aforementioned problems that expert-ethicist centred analysis brings. A top-down matrix cannot support deliberation in this capacity. If the supposedly 'correct' values are prescribed prior to the engagement process (including the inherent Western philosophical bias of the pre-defined principles) then the bottom-up nature of deliberation is removed. With this in mind, bottom-up deliberation with participant ascribed principles is required. In spite of this, four problems remain.

Firstly, although it is argued here that bottom-up principle and stakeholder selection is appropriate, a further tool is necessary in order to achieve this in a transparent and meta-ethically justified manner. Secondly, the range of principles and stakeholders offered by the matrix is comparatively small. A far greater range of stakeholders and principles would be needed to alleviate the inherent bias in the model generated by such a small selection. The identification and display of such a narrow set of principled perspectives and stakeholder groups could have two outcomes. It may lead to participant conflict over those groups that were chosen to be included in the matrix and those that weren't – a problem that may simply lead to a redrawing of a larger matrix with more representative groups. More significantly, important stakeholder groups absent from the matrix may be overlooked because they were not on the deliberative agenda, thus precluding them from informed discussion. Thirdly, the matrix structure frames the deliberative agenda through inclusion and exclusion of certain groups. Thus, a meta-ethically justified process for the selection of stakeholders is necessary - a mapping device for identifying actors and the relationships between them. This process may take longer than simply making the matrix much bigger and spending the extra time filling in all the cells, although the selection of these inputs to the matrix is itself a deliberative process that requires structure, and hence deserves a facilitation tool in its own right. Fourthly, the matrix in its current form also lacks suitable deliberative mechanisms for closing-down ethical decision-support processes. In a closing-down phase the aim is to instrumentally assist policy making by, "cutting through the messy, intractable and conflict-prone diversity of interests and perspectives to develop a clear, authoritative, prescriptive recommendation to inform decisions" (Stirling 2004).

In a revision of the matrix, Mepham includes a weighting mechanism for ethical evaluation, separating positive and negative ethical impacts where a score is applied according to whether the principle is respected or infringed; weighted by scoring along a Likert-type scale, i.e. -2 (strongly infringe a principle) to +2 (strongly respect a principle). He argues that scoring perceived ethical impacts on a numerical scale may serve as a means of establishing relative perceptions, but the framework should not be viewed as a decision model. Indeed as Whiting (2004) argues, "...depending on the weighting given to various cells in the matrix almost any ethical evaluation can be supported." Schroeder and Palmer (2003) highlight that simply counting the numbers of infringed and upheld principles has in itself an inherent utilitarian bias (thus procedures like the one Gamborg [2002] suggests, inherently prioritise the ethical values of the many over the few). Also weighting criteria based upon a hierarchy of principles are equally problematic as they contradict pluralistic ethical deliberation by arbitrarily prioritising certain ethical principles over others. In the absence of reliable weighting criteria and hence a closing down mechanism for evaluation, summary and prescription, ethical decision-making remains reliant upon the competency of the users' moral judgement, so greater clarification and structured deliberation around conflicting moral judgements is necessary.

4.4   The Ethical Grid

The second ethical tool, labelled the Ethical Grid (EG), is presented in David Seedhouse's book 'Ethics: the Heart of Health Care' (Seedhouse 1988; Seedhouse 1998). Seedhouse argues that the abstract philosophy presented by Western normative ethics is largely inappropriate for the decision-making realities of healthcare practice. The EG is designed to provide down-to-earth guidance for individuals to analyse ethical problems for themselves (Seedhouse 1998) and is presented as an ethical 'tool' in a fairly literal sense. Seedhouse uses the analogy of a spade; like a good gardener the 'grid user' understands the importance of keeping the tools clean and sharp, and understands when it is appropriate to use it (ibid). Elsewhere he states that "like a hammer or screwdriver used competently, it can help make certain tasks easier, but it cannot direct the tasks nor can it help decide which tasks are the most important. The grid can enhance deliberation – it can throw light into unseen corners and can suggest new avenues of thought – but it is not a substitute for personal judgement" (Seedhouse 1998).

The theoretical basis of Seedhouse's work divides the ethical realm into two distinct forms, which he labels Ethical A and Ethical B. Ethical A means ethical in the sense of having ethical content, and Ethical B in the sense of having a consistent view about what one ought to do in the social world. The Ethical A position appears to be grounded in a 'negative liberty' conception of ethics in the social world, whereby, as Thomas Hobbes (1651/1998) argued, "a free man is he that... is not hindered to do what he hath the will to do". Or which, as Isaiah Berlin argued in his 'Two Concepts of Liberty' essay, involves answering the question,

"What is the area within which the subject - a person or group of persons - is or should be left to do or be what he is able to do or be, without interference by other persons." (Berlin 2002). Seedhouse applies negative liberty to ethics. In one analogy he highlights how actions such as twirling one's hair or tapping on a desk is of no moral importance unless it is in a shared office and thus interferes with the work of others (Seedhouse 1998). The world of 'Ethical A' is defined socially - characterised by a complex world that is continually fraught with ethical dilemmas that require resolution by the actors that inhabit it. This contrasts with 'Ethical B' as it is by definition the realm of normative ethics; concerned with how an individual ought to act in social interrelationships. In the rather trivial case of someone tapping a desk, the move from Ethical A to Ethical B involves the realisation that such actions are irritating and thus influential. Unnecessarily irritating others at work is unfair to them, and therefore one's 'duty' to stop is normatively motivated. 'The ethical' is, to Seedhouse, intrinsically linked to social interrelationships of individuals. The move from existing in a social world full of ethical dilemmas to participating in ethical problem solving and decision-making, therefore requires reflection about those everyday interactions between individuals.

The repeated reference to practical and everyday analogies in illustrating the realm of the ethical serves to underpin Seedhouse's assertion that ethical behaviour is part of everyday existence and interaction; that it is not a sterile academic pursuit or thought experiment, but is in fact an intrinsic aspect of the conduct of everyday professional practice, as abstract and contested concepts of ethics lead to individuals ignoring or dismissing ethical issues and conflicts. Thus, the agenda of the EG is to allow (or perhaps more accurately persuade) health care professionals such as doctors, nurses, social workers etc, to take command of the realm of Ethical A by committing to the model of Ethical B, illustrated in the EG itself.

Seedhouse identifies a realm of ethics whereby the complexity of the moral world remains largely hidden from view; individuals perceive merely the 'tip of the (moral) iceberg' (1998) as Seedhouse puts it. The actor must stand upon the tip of this 'iceberg' constructed by Ethical A; at any given time or within any particular context only a portion of the full ethical issue is on view. One could argue that this is a critical realist, deep ontology of ethics; that requires the individual to maintain a reflexive understanding of ethical practice in a world where moral complexity cannot be fully observed. In answer to this, the grid represents a tool that allows the individual to uncover more of this complex ethical world and thus act to achieve the normative goals that are consistent with this social/ethical realm. The fundamental focus then becomes the idea of 'doing' ethics, through reflection, reasoning and application in everyday practice. It thus becomes practical ethics rather than applied ethics. The EG itself is designed primarily as a practical and visual tool that allows the practitioner to manipulate and reflect upon the issues presented in (seemingly) logical and rational order. Visually, the square grid divides the ethical concepts within into twenty boxes using concentric rings and bisecting lines. Each box contains a single ethical concept, so individual boxes can be self-contained and detachable, in the manner shown in figure 4.1 derived from (Seedhouse 1998).

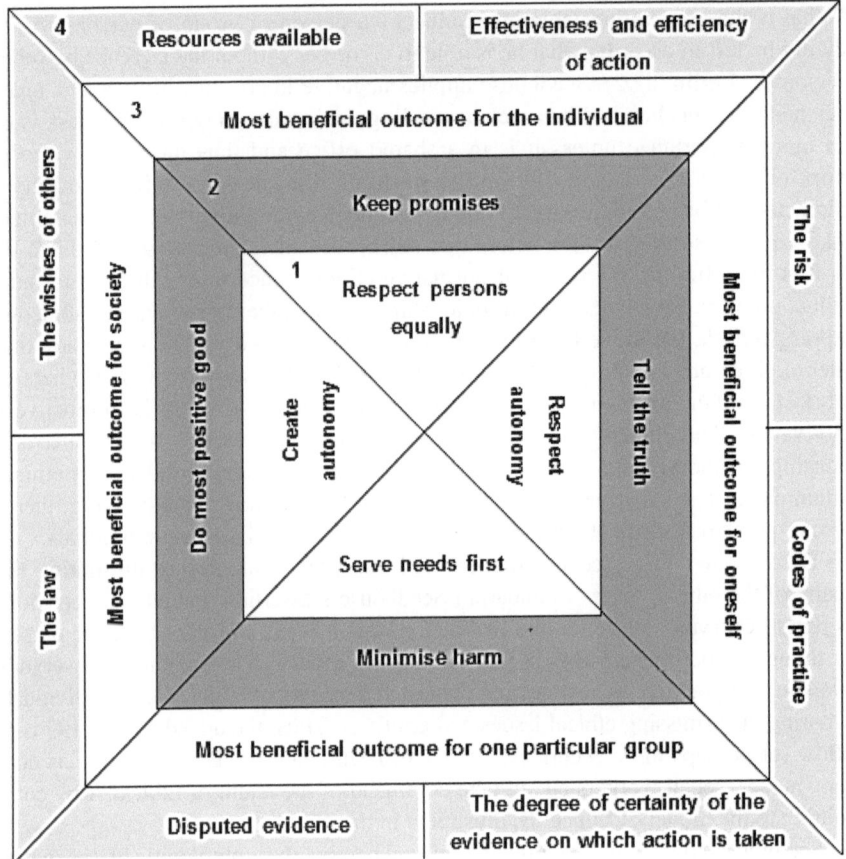

Fig. 4.1 The Ethical Grid

The EG is essentially an artificial device, so the separation of the concepts contained within each box is therefore also artificial. Seedhouse proposes that the implementation of the EG can take a number of different forms. It could be used to take each box in order and complete the whole grid in a set order, or else one could start at the centre and spiral outwards, with the central boxes as the most important the outer boxes the least important. Alternatively the most important from each layer could be used. The implication from this is that the grid is presented as a holistic and more importantly, a complete set of ethical principles to apply to health care. The fact that Seedhouse advocates a number of different processes for using the EG with no fixed order implies that the grid is presented as a robust and adaptable tool to apply in a variety of contexts. The visualisation of the EG is also flexible. It can be imagined as either a two or three dimensional construct according to the will of the user. It can be visualised as a four sided pyramid (or constructed as such as a three-dimensional model), and thus each side can be considered in turn, although Seedhouse asserts that there is no special relationship between the boxes on any one side. Alternatively, in two dimensions it could be envisioned as "written on a piece of rubber", with "an invisible cord at its centre", which can, "pull the Grid either towards or away from the viewer" in order to keep the EG in mind and in view as a whole (Seedhouse 1998).

The potential visualisations of the grid allow a degree of flexibility, although this also creates a problem with regards to the complex task of recording how an ethical decision is made. To use a paper-and-pencil diagrammatic approach to the grid would require a constant redrawing and reshaping of the grid's structure (which is inconvenient, impractical and would tend to dull the analytical sharpness of the grid as a tool). Similarly, physically constructed three-dimensional grids from wood, card, plastic etc, are flexible but cannot be recovered, i.e. physically recorded (which is an important aspect of deliberative methods, in that they should be explicable to third party representatives). Computer aided input and visual manipulation of a software-based grid could potentially alleviate this process (Machlaren 2001), and much of Seedhouse's recent work on the EG has been to this effect. Computer mediated ethical decision-making has a number of possible merits in terms of resolving the practicality/recoverability problem. However, group deliberation is not always conducive to systematic data input into a computer system, particularly when simpler paper-and-pencil techniques are more readily available.

4.4.1 The Normative Theoretical Underpinnings of the Ethical Grid

As shown in figure 4.1, the grid is a diagrammatic structure composed of four boxed layers, usually presented in different colours, though labelled here as 1-4. Its four coloured layers are used to illustrate the different approaches in conceiving what is ethical, by the dominant theories of normative ethics. Layer 1 (normally blue), the core of the EG, represents the rationale of Seedhouse's conception of an ethical health care system and the normative basis upon which to conceptualise the meaning of 'working for health'. What is interesting about layer 1 is that it represents a prescriptive central point – a set of primary principles upon which the foundation of Seedhouse's conception of ethics stems; the cultural/moral framework by which he wishes to reconstruct the institution of the health service. This seems somewhat ironic given the vehement attack made that he makes on the principlist approach to medical ethics presented by Beauchamp and Childress.

Seedhouse critiques principlism as being adopted almost as if it were a mantra within the health care sector; individuals citing the four principles without critical reflection on behalf of the actors utilising such an approach. It is interesting that, given this rejection of principlism as being "nebulous" and as being "generally acceptable to well-heeled Western liberals [who] do no more than offer conclusions (a) open to wide interpretation and (b) acceptable only to those who agree with them in the first place" (Seedhouse 1998), Seedhouse then sees fit to place at the centre of his grid a layer of central principles "at least one [of which] must be used during deliberations" (ibid.: 39). It seems apparent therefore that a central set of core principles is unavoidable within ethical decision making contexts, and the grid does not allow for the complete removal of a principle driven approach without "massive justification" (ibid.: 39). As in the case of the ethical matrix, if adapting the grid for use in the PTA context these central principles should be examined and kept or rejected on the basis of their relevance to the case in point (i.e. facility siting in a particular community), and to the overall internal coherence of the grid itself. Changing one set of principles may involve the augmentation or substitution of some or all of the boxes within the grid. Layer 2 (normally red) is representative of duties and motives. Its fundamental basis is upon the deontological position in normative ethical theory. The significance of this layer is to elicit the duties and obligations that are "implied by a commitment to health work" (ibid.: 42):

- Keep promises
- Tell the truth
- Minimise Harm
- Do most positive good

Seedhouse does not suggest that these duties are the only ones possible or that they should be binding, but argues against removing these obligations unless suitable justification is supplied. It would also be possible to supplement the duties presented for others, but similarly justification for the choice of different duties over those originally specified should be supplied. The difficulty that this presents is that the act of choosing the correct four duties to uphold in any given technology context (and whether it is significant to choose four in the first place) would itself be a complex deliberative process, something which will be examined further later in the chapter.

Layer 3 (normally green) is based upon a utilitarian perspective contrasting to the deontological grounding of the preceding layer 2. The utilitarian context is neatly (and rather conveniently it again appears) divided into 4 subcategories:

1. Most beneficial outcome for the individual
2. Most beneficial outcome for oneself
3. Most beneficial outcome for a particular group
4. Most beneficial outcome for society

Here the idea of 'most beneficial outcome' is applied to the differing layers of analysis: 'the individual', 'the group', or 'society' levels. This appears to equate roughly with examining the micro, meso and macro levels of the consequences of ethical action. The focus here is about setting health care commitments in terms of priorities, and the consideration of all affected parties. However such an analysis of grouping the affected parties into what effectively amounts to 'you/me', 'them', or 'all of us' would be rather simplistic for transferring to any participatory technology assessment context. It is widely recognised that priority setting is a contentious and politically fraught process, reflecting the complex interests and interactions of stakeholders (including intangible interests such as those of 'future generations' or 'the environment'). Seedhouse's somewhat over-simplified model of affected parties would appear to bypass deliberation in favour of simple categorisation and hence is an inappropriate basis upon which to analyse the complex stakeholder and community relationships and interactions to be found the assessment of controversial technlogies.

The final layer 4 of the EG is the realm of external considerations, those factors outside of the moral sphere that necessarily influence or impinge on decision-making. This layer includes the legal, political, social and wider institutional and interpersonal factors which, although they may all have a moral component, are still external to the ethical decision making process; in effect the 'independent variables' in the decision. This is important as decision-making on ethical grounds may be constrained by ethically neutral or non-negotiable factors such as regulatory mechanisms, policies and statutes. The negotiation between ethical and non-ethical factors is therefore worthy of deliberative evaluation in itself.

4.4.2 Critique of the Ethical Grid

One of the primary criticisms of the grid is simply its constraint within four specific normative theoretical traditions. Like the ethical matrix, the four sided structure helps to simplify the model (conceptually and visually) but also severely limits the choice of principles and perspectives examined. The limit of four principles, four duties, four outcomes and eight external influences, places normative restrictions upon the user. Seedhouse's selection of ethical criteria is essentially arbitrary, and relates to the stated intention to change the health care sector. The grid constrains the ethics of health care practice within a specific normative framework. The choice of principles is limited (in the first instance) to those that the originator deems relevant. Arguably the principles are suitable for the context in which they are supplied (i.e. in the face of a health care system that is moving towards a 'management model' of output delivery and patient services supplied as 'goods' rather than holistic patient care) and Seedhouse does leave the grid open to a different selection of principles (given adequate meta-ethical justification for doing so). However, unless the choice of ethical principles is meta-ethically justified by the practitioners themselves, the grid is not a 'bottom-up' ethical tool, in the sense that the normative ethical content is pre-defined (by the grid designer, even if principles are substituted for others) rather than user-defined, creating what is termed a 'framing effect' within the decision, which is in essence top-down.

One possible solution would be simply to leave the grid blank initially and use other forms of deliberative procedure to elicit the principles, duties, outcomes and externalities to be included. To do this, however, requires normative weighting or prioritising, highlighting a characteristic practical problem of this type of applied ethics. At some stage, the complexity of ethical inquiry must be reduced in order to create a tool which is 'sharp' and efficient. There is an inevitable trade-off between complexity and analytical depth on one hand, and simplicity and ease of use on the other.

One might also take issue with the 'squareness' of the 4,4,4,8 grid format. Such a format does not reflect a natural internal consistency between the ethical principles presented, nor is it a formation that covers a full range of ethical positions; it is merely a diagrammatic shape into which concepts are arranged. This appears to be a design consideration for user simplicity rather than based upon philosophical justification. One might also question whether the numbers of spaces presented for each ethical theory category (deontology, consequentialism, etc...) is sufficient or whether more or less are called for, as no specific justification is presented in Seedhouse's work. For example, a triangular, pentagonal or even dodecahedronal shape may be more appropriate, if more principles are considered relevant to the case in hand.

In short, the ethical grid has a number of features that are useful to the design of ethical tools for participatory technology assessment. The EG is designed primarily for individual users to reflect upon their behaviours in relation to broader ethical concepts and dilemmas. The divisions between Seedhouse's Ethical A and Ethical B show an interrelationship between the idea of act-deontology (whereby individuals must choose morally valid courses of actions) and an ontologically deep moral social world (where the ethical consequences of actions cannot be fully understood by individuals). Moral actors must also make decisions within the constraints of external social forces such as regulation, policy and law. In many respects this shares a commonality of approach with the aforementioned wide reflective equilibrium approach and thus deserves some consideration for application as a decision-support tool. The application of a tool designed primarily for use by individual practitioners in specific medical scenarios to a group-based decision-support process presents a number of challenges, however.

The EG was designed for health care practitioners dealing with individual patients. It therefore allows users to consider different theoretical perspectives and external considerations, but is not deliberative in the sense that it allows opportunity for group discussion and engagement around theory or issue selection. That said, the grid has design facets, notably its categorisation and colour-coding of ethical constructs, which may usefully inform further development of deliberative ethical tools. However, non-specialist participants with no formal ethics training may struggle to wield such conceptually weighty notions as act-deontology, utilitarianism or consequentialism. Simplifying the language and dividing and colour-coding ethical positions would seem likely therefore to be beneficial for facilitating ethics deliberation among groups of non-specialist citizens or stakeholders.

4.5 Ethical Delphi

The third tool that I consider is called the Ethical Delphi. It is an augmentation of the Delphi method developed within the RAND Corporation in the 1950s by researchers involved in a US Air Force project. Their original aim was the application of expert opinion to the selection – from the point of view of a Soviet strategic planner – of an optimal U.S. industrial target system, with a corresponding estimation of the number of atomic bombs required to reduce munitions output by a prescribed amount (Rowe and Wright 1999). The history of the method is therefore steeped in technocratic assessment and decision-making. Since the 1950s the Delphi method has re-emerged as a method to combine knowledge and abilities of a diverse group of experts to evaluate developments that were deemed to lie outside traditional scientific assessments; either because decisions involved elements of judgement or dealt with uncertainties of various kinds (Millar 2007; Stewart 1987) and has been applied to a great range of decision and policy-making contexts (Hasson et al. 2000). It has been used for generating ideas, exploring future scenarios, collecting data and supporting decision-making in a range of contexts, from energy planning, to healthcare and social policy; and has been used to assist policy-making processes when information is incomplete or the validity of that information is in dispute (Adler and Ziglio 1996; Hill and Fowles 1975; Rauch 1979).

In practice, a Delphi obtains forecasts from a panel of independent experts over a period of two or more 'rounds' during which experts are asked to predict quantities. After each round, an administrator provides an anonymous summary of the experts' forecasts and their reasons for them. The process reaches a 'saturation point' when experts' forecasts have changed little between rounds. At saturation the process is stopped and the rounds are averaged. Proponents of the Delphi model suggest that it is based on well-researched principles and provides forecasts that are more accurate than those from unstructured groups (Rowe and Wright 1999, 2001). A key feature is the anonymity of the expert participants. The Delphi is structured around a virtual committee with anonymous and remote exchange of ideas, values and positions through a series of opinion exchanges. The participants convene as an ad hoc committee that will communicate remotely (most commonly now through electronic media) without ever meeting in person. In each successive round, the panel of participants is asked to comment upon the values and viewpoints expressed in the previous opinion exchange. The Ethical Delphi essentially elicits ethical issues, judgements and considerations that the expert panel deems are relevant and significant for the decision process. Where the Ethical Delphi and the classical Delphi method diverge, is that the ethical version does not seek overall consensus from its participants for future action or development, but instead serves to highlight areas of consensus and disagreement between participants and then map these to show the diversity of ethical values involved in complex technological decisions (Millar 2007; Millar et al. 2006). Millar et al (2007; 2006) define the context in which the Ethical Delphi approach would be useful:

- Expert input is required for policies under review or development
- Issues are uncertain, controversial and complex
- Judgement and weighing of arguments is essential
- Many and diverse research communities and stakeholders have concerns
- Outcomes from the process should have an impact on several issues, including future policy making
- There is need for a cross-sectoral scientific debate

The Ethical Delphi highlights the idea that deliberative methods should not be based on consensus-building alone. Indeed one must question if consensus is necessary in ethical deliberation at all. Given the plurality of values and ethical positions that may exist around an issue, consensus (especially in small, statistically non-representative groups) may serve little value. As I argue in the previous chapter, the consensus of ethical opinion within a group of experts cannot represent the diversity of public ethical values, and so to rely upon such consensus building within an ethical Delphi approach could again be criticised for being ethically technocratic.

4.5.1 A Critique of the Ethical Delphi

The first critique of the Ethical Delphi concerns the expert driven nature of the approach. The Delphi technique originated in the 1950's as a process for eliciting diverse expert opinions and weighting them to produce a consensual decision among those with competing or contrasting scientific and technical interests. In one sense it is a deliberative tool, although with sole input from experts and anonymity throughout the process, two key issues arise. Firstly, decision outcomes lack social 'robustness' in terms of how well they represent affected community groups or stakeholders. Secondly, outcomes are neither open nor transparent if decisions are made in a 'black box' away from public scrutiny.

The idea of having a series of rounds with anonymous input was designed originally to reduce the possibility of bias and political coercion within the discussion setting, principles that would be consonant with those which Habermas propounds as the basis of discourse ethics. In some respects this is unnecessary in the modern deliberative policy-making context. Stakeholders and communities under the deliberative turn, are openly and actively engaged in networks of interrelated communication practices about the issues under consideration. Although one advantage of the Delphi is that is can be done remotely, via post or electronically, the anonymity aspect is questionable in a well established expert and stakeholder network such as that which often exists around new technologies, where many stakeholders may be able to recognise the values and judgements expressed as belonging to one another due to the familiarity that results from sustained mutual engagement on these issues. Also within this research, the key goal is the active engagement with communities on their diverse values. As previously argued, scientists and ethical specialists have no particular moral authority or insight that differentiates their values as superior to those of affected stakeholders or citizens. Therefore the elicitation of expert opinion runs contrary to one of the central arguments of this book. Although the Ethical Delphi method has promise in broadening out purely technical and scientific debates to a greater level of values and viewpoints, it lacks the openness, transparency, and public-centred legitimacy recognised to be appropriate and necessary in this case.

4.6 Conclusions

The foregoing discussion has established that for a variety of reasons, the adoption of these three ethical tools is inappropriate for participatory technology assessment contexts. Other criticisms are philosophical. In both the ethical matrix and the ethical grid, questions are raised over the choices of the principles used; what, if anything, can justify the choice of the three or four principles presented over others? The answer is grounded in part by practical necessity, the inclusion of too many ethical principles or theory perspectives increases complexity and slows completion of the ethical assessment. Transposing the ethical matrix or grid from bioethics or health care ethics to new technology contexts (such as nuclear power or synthetic biology) would require a reassessment of the selected principles or theory perspectives. However, selecting specific principles over others requires meta-ethical justification, which these tools alone do not provide. As previously stated, the ethical principles employed are grounded in three dominant strands of what could be loosely termed Western philosophy: utilitarian consequentialism, Kantian deontology, and Rawlsian contract theory. Although accepting these principles may be justified on the basis of their familiarity for participants, designing new tools to incorporate a broader selection of principles or theoretical perspectives has the advantage of widening ethical assessment and making it more relevant to the case in hand and the differing perspectives of the stakeholders involved. The task is therefore to develop a tool that can cope with a greater breadth of ethical principles, without dulling the effectiveness of the tool in practical terms.

The primary concern with the both the EM and EG is the way in which the shape each tool constrains deliberation. In the grid for example, what value does constructing the tool as a series of layers in a bisected square add to the conceptual model used to tap into the complex world of social ethics? Seedhouse's original concept for an ethical tool was an interdependent web of ethical possibilities, with different regions and seemingly infinite routes through which one can travel to reach ethical conclusions (Seedhouse 1988). This was ultimately rejected because despite its elegance, it was deemed too complex and daunting for his students (and practitioners more generally) to use effectively. Nevertheless, Seedhouse's original vision is intriguing and suggestive of other possibilities. The problem of complexity and his reservations about such an interrelated approach may be overcome to a great extent if one turns to ethical deliberation in a group context. There already exists a wide repertoire of deliberative tools and techniques used to facilitate group interaction and co-operative problem solving. By building upon these existing tools and by using an iterative research design to test and reshape the process in light of experience, a pragmatic trial-and-error method may yield a more elegant and complex system of ethical deliberation. An approach such as this may better reflect Seedhouse's insight into the interrelated nature of the ethical world in a simpler and more transparent way. The remaining chapters within this section report upon the development of such a model, and the final section of this book illustrates its application.

References

Adler, M., Ziglio, E.: Gazing into the Oracle: The Delphi Method and its application to social policy and public health. Jessica Kinglsey Publishers, London (1996)

Beauchamp, T.L., Childress, J.F.: Principles of Biomedical Ethics, 5th edn Aufl. Oxford University Press, New York (2001)

Beekman, V., Brom, F.W.A.: Ethical Tools to Support Systematic Public Deliberations About the Ethical Aspects of Agricultural Biotechnologies. Journal of Agricultural and Environmental Ethics 20(1), 3–12 (2007)

Berlin, I.: Liberty (revised and expanded edition of Four Essays on Liberty). Oxford University Press, Oxford (2002)

Boucher, P., Gough, C.: Mapping the ethical landscape of carbon capture and storage. Poiesis & Praxis 9(3-4), 249–270 (2012)

Chadwick, R., Henson, S., Moseley, B., Koenen, G., Liakopoulos, M., Midden, C., Palou, A., Rechkemmer, G., Schröder, D., Wright, A.V.: Functional Foods. In: Ethics of Science and Technology Assessment. Springer, London (2003)

Cotton, M.: Evaluating the 'ethical matrix' as a radioactive waste management deliberative decision-support tool. Environmental Values 18(2), 153–176 (2009)

FEC. Ethical Matrix: Uses (2005),
http://www.foodethicscouncil.org/ourwork/tools/ethicalmatrix/uses (accessed February 02, 2007)

Forsberg, E.M., Kaiser, M.: Ethical decision making in STRATEGY - final report and recommendations. National Committee for Research Ethics in Science and Technology (NENT), Oslo (2002)

Gamborg, C.: The acceptability of forest management practices: an analysis of ethical accounting and the ethical matrix. Forest Policy and Economics 4, 175–186 (2002)

Hasson, F., Keeney, S., McKenna, H.: Research Guidelines for the Delphi Survey Technique. Journal of Advanced Nursing 32, 1008–1015 (2000)

Hill, K.Q., Fowles, J.: Delphi forecasting technique. Technological Forecasting and Social Change 7, 179–192 (1975)

Hobbes, T.: Leviathan. Oxford University Press, Oxford (1651/1998)

Howard, B.J., Forsberg, E.M., Kaiser, M., Oughton, D.: An Ethical Dimension to Sustainable Restoration and Long-Term Management of Contaminated Areas. In: International Conference on Radioactivity in The Environment, Monaco (2002)

Kaiser, M.: Xenotransplantation-ethical considerations based on human and societal perspectives. Acta Veterinaria Scandinavica Supplementum 99, 65–74 (2004)

Kaiser, M., Forsberg, E.M.: Assessing Fisheries - Using an Ethical Matrix in Participatory Processes. Journal of Agricultural and Environmental Ethics 14, 191–200 (2001)

Kaiser, M., Millar, K., Forsberg, E.-M., Baune, O., Mepham, B., Thorstensen, E., Tomkins, S.: Decision-Making Frameworks. In: Beekman, V. (ed.) Evaluation of Ethical Biotechnology Assessment Tools for Agriculture and Food Production: Interim Report Ethical Bio-TA Tools, Agricultural Economics Research Institute, The Hague (2004)

Kaiser, M., Millar, K., Forsberg, E.M., Thorstensen, E., Tomkins, S.: Developing the ethical matrix as a decision support framework: GM fish as a case study. Journal of Agricultural and Environmental Ethics 20(1), 53–63 (2007)

Machlaren, P., Seedhouse, D.: Computer Mediated Communication with Integrated Graphical Tools Used For Health Care Decision-Making (2001)

Mepham, B.: Ethical analysis of food biotechnologies: an evaluative framework. In: Mepham, B. (ed.) Food Ethics, pp. 101–119. Routledge, London (1996)

Mepham, B.: The role of ethics in food policy. Proceedings of the Nutrition Society 59, 609–618 (2000)

Mepham, B.: A Framework for the Ethical Analysis of Novel Foods: The Ethical Matrix. Journal of Agricultural and Environmental Ethics 12, 165–176 (1999)

Mepham, B.: Novel foods. In: Chadwick, R. (ed.) Concise Encyclopedia of Ethics and New Technologies, pp. 300–313. Academic Press, San Diego (2001)

Mepham, B.: A framework for ethical analysis. In: Mepham, B. (ed.) Bioethics: An Introduction for the Biosciences, Oxford University Press, Oxford (2005)

Mepham, B., Tomkins, S.: Ethics and Animal Farming: A Web-based interactive exercise for students using the Ethical Matrix (2003), http://www.ethicalmatrix.net (accessed October 21, 2010)

Millar, K., Thorstensen, E., Tomkins, S., Mepham, B., Kaiser, M.: Developing the Ethical Delphi. Journal of Agricultural and Environmental Ethics 20, 53–63 (2007)

Millar, K., Tomkins, S., Thorstensen, E., Mepham, B., Kaiser, M.: Ethical Delphi Manual. LEI, The Hague (2006)

Oughton, D., Bay, I., Forsberg, E.M., Kaiser, M.: Value Judgements and Trade-Offs in Management of Nuclear Accidents: Using an Ethical Matrix in Practical Decision-Making. In: VALues in Decisions on Risk. Stockholm (2003)

Oughton, D.H., Bay, I., Forsberg, E.M., Hunt, J., Kaiser, M.: Social and Ethical Aspects of Countermeasure Evaluation and Selection - Using an Ethical Matrix in Participatory Decision Making. Deliverable Report of the STRATEGY project, Oslo (2003)

Rauch, W.: The Decision Delphi. Technological Forecasting and Social Change 15, 159–169 (1979)

Rawls, J.: Outline of a Decision Procedure for Ethics. The Philosophical Review 60(2), 177–197 (1951)

Rawls, J.: A Theory of Justice, 2nd Aufl. Oxford University Press, Oxford (1999)

Rowe, G., Wright, G.: The Delphi technique as a forecasting tool: issues and analysis. International Journal of Forcasting 15, 353–375 (1999)

Rowe, G., Wright, G.: Expert opinions in Forecasting: The role of the Delphi technique. In: Principles of Forecasting: A Handbook for Researchers and Students. Kluwer Academic, Boston (2001)

Schroeder, D., Palmer, C.: Technology Assessment and the "Ethical Matrix". Poiesis & Praxis 1(4), 295–307 (2003)

Seedhouse, D.: Ethics: the Heart of Health Care. Wiley, Chichester (1998)

Seedhouse, D.: Ethics: the Heart of Health Care. John Wiley and Sons Ltd., Chichester (1988)

Small, B.H., Fisher, M.W.: Measuring Biotechnology Employees' Ethical Attitudes towards a Controversial Transgenic Cattle Project: The Ethical Valence Matrix. Journal of Agricultural and Environmental Ethics 18(5), 495–508 (2005)

Stewart, T.R.: The Delphi Technique and Judgemental Forecasting. Climatic Change 11, 97–113 (1987)

Stirling, A.: Opening Up or Closing Down? Analysis, participation and power in the social appraisal of technology. In: Leach, M., Scoones, I., Wynne, B. (eds.) Science, Citizenship and Globalisation. Zed, London (2004)

Whiting, T.L.: Application of the ethical matrix in evaluation of the question of downer cattle transport. In: Can West Veterinary Conference, October 2-5. The Banff Centre, Alberta and British Columbia Veterinary Medical Associations, Banff, Alberta Canada (2004)

Chapter 5
Reflective Ethical Mapping

5.1 Introduction

In the previous chapter I outlined a series of specifically formulated methodological tools to encourage reflection and deliberation on ethical issues in a real world decision-making context. My critique of these approaches has been both philosophical and practical in nature. In the case of the Ethical Matrix (EM) and Ethical Grid (EG) the constraints are based primarily upon their format. Matrix and grid structures inhibit the identification of a broad range of relevant public actors, stakeholders, environments, ethical principles and socio-technical concerns because these methods limit both the quantity of such factors for discussion, and in some cases, constrain the choice of these elements without sufficient meta-ethical justification.

What I propose in this chapter is to set out a practical model for ethical PTA that expands upon these pre-existing methods by opening them up to more effective bottom-up deliberation on these important elements. The EM and EG lack sufficient opening-up and closing-down mechanisms to first elicit ethical reflection by individual stakeholders upon a range of socio-technical issues and relevant ethical perspectives through open and lively discussion, and then later to bring these discussions to a decision-point, summary and evaluation. To be effective it is important to expand ethical deliberation to include not just the assessment of principles and stakeholder interests to specific cases, but also the justification of choices when selecting principles and stakeholders within cases. Crucially, as a point of pragmatic consideration, it is necessary to also relate ethical deliberation back to a specific policy context, and thus make it practically useful to PTA decision-making. I propose in this chapter that multiple ethical tools are necessary, not simply in the format of a tool box (implying different tools brought out for different purposes) but arranged as a sequential decision-support procedure, to better satisfy the opening-up and closing-down mechanisms necessary within PTA. In the following chapters I present a series of practical methods structured in sequence as a decision-making procedure. In each of the methods chapters I include some discussion of empirical examples drawn from participatory work surrounding the issues of technology development and decision-making in the decommissioning of nuclear power facilities and the long term management of radioactive wastes.

M. Cotton, *Ethics and Technology Assessment: A Participatory Approach*,
Studies in Applied Philosophy, Epistemology and Rational Ethics 13,
DOI: 10.1007/978-3-642-45088-4_5, © Springer-Verlag Berlin Heidelberg 2014

5.2   The Features of Deliberative Decision-Making Methods

When beginning to develop novel participatory-deliberative decision-making tools it is important to begin with an understanding the nature and sequences of decision-making, both regarding ethical and non-ethical decisions; and the ways in which deliberative processes are structured in order to achieve specific goals. The first point of consideration is that decision-making processes are frequently multi-staged, sequentially constructed and iterative in nature. This is especially true of participatory-deliberative decision-making processes which are, in essence, problem-solving activities. They require participants to identify and then define a series of questions, difficulties and challenges pertinent to the problem situation in hand. They usually begin by collaborative identification and discussion of potential solutions whilst explaining and mutually challenging the underlying reasoning behind such solutions. Participants must then verify, accept or reject such solutions based upon some mechanism of evaluation. This evaluation is usually based upon a predefined criterion or metric against which to assess outcomes. The types of decision-making that commonly occur within PTA are usually formulated as multi-staged processes in this manner. They typically involve a searching phase to discover goals, followed by the formulation of objectives, selection amongst alternative options and then the formulation of strategies to accomplish the objectives and an evaluation of outcomes. Thus as a problem-solving exercise, participatory-deliberative decision-making in Technology Assessment follows a familiar pattern:

- Recognise a problem
- Identify a series of objectives
- Collect information and ideas
- Analyse the information and ideas
- Choose a specific course of action, i.e. make a decision
- Communicate and implement the decision
- Assess the outcome
- Evaluate and report on the outcome
- Recognise new problems and repeat the process where necessary

The literatures on deliberative tools to achieve PTA goals are rapidly growing both in the academic and policy literatures, and many such as Multi Criteria Decision Analysis (Nijkamp 1989), Multi-Criteria Mapping (Stirling 2001), Stakeholder Decision Analysis (Burgess 2006) and the hybrid Deliberative Mapping (Burgess 2007) share this type of structure. I propose that an ethical tool-based approach could benefit from the adaptation of the methods examined in the previous chapter in line with this model; separating into individual methods for each of the sequential stages.

5.3 Ethical Decision-Making

What is true of problem solving and participatory-deliberative decision making is, for the most part, also true of ethical decision-making. One common feature of the applied ethics literatures (especially common in the field of professional ethics such as medical or business ethics) is the development of ethical decision-making procedures. Broadly speaking, many of the decision-making frameworks that have emerged in business, healthcare, engineering and other professions, are commonly grounded within the act-deontology tradition of ethics: centred upon the role of the individual actor making a decision laden with ethical consequences and charged with making the right choice whilst considering a range of outcomes. Many methods of ethical decision-making are designed to facilitate and structure this process of moral reflection in order to consider different inputs, perspectives and eventualities before coming to a decision on how to proceed with a course of action.

Rather than detail all the available ethical decision-making models individually, it is clear that many of the models that have emerged within the applied ethics literatures share a common step-wise structure. Ethical decision models tends to be patterned in a checklist, decision-tree or similar sequential model, whereby the individual actor moves through a series of evaluation stages in order to reach a better informed and ethically robust decision at the end. Most of these models involve assessing relevant information followed by normative theoretically informed reflection that influences the moral actor in reaching a decision. Some recognise that by completing and evaluating the ethical implications of action, this in turn presents a self-perpetuating hermeneutic cycle. As one decision closes this raises new ethical questions for consideration and hence further rounds of questioning, exploring consequences and reaching conclusions. To give a basic overview, I suggest a general format of checklist approaches that can be broadly summarised in this way[1]:

1. Recognise an ethical question, issue or concern
2. Assess the relevant facts and values
3. Evaluate alternative actions from theoretical perspectives
4. Weight outcomes on the basis of ethical perspectives
5. Implement the decision outcome
6. Reflect upon and evaluate the outcome
7. Repeat step 1 as appropriate

A number of available ethical decision models can be found in the applied ethics literature, many of which are implicitly or explicitly grounded in the act-deontology tradition, holding that moral judgements are particular and case specific, tending to look towards the consequences of specific decisions in terms of whether that are beneficial or harmful, and then acting accordingly to what is

---

[1] For further details of each decision-making model, refer to (ERC 2004; Thomson 1999; Marshall 1999; Forester- Miler and Davis 1996; Van-Hoose 1980; Bowen 2005; Potter 1999; Jones 1991).

deemed to be right by the morally situated actor. Though care must be taken when generalising across all forms of ethical decision-making model, the preponderance with act-deotonlogy has emerged principally because much of the applied ethics literature tends to focus upon actor decision-making, particularly in specific professions such as business management, engineering or medicine. It also remains concerned with the ethical implications of personal action to other individuals (business owners, citizens, patients etc.) and the focus is largely upon the implications of an individual's actions within and amongst broader professional organisations.

In the context of this book, however, the notion of an ethical decision-making model incorporates two additional criteria against which an act-deontological model must contend. The first is that by focussing upon the individual's actions and behaviours, these models tend to bracket out the role of technologies, which as already mentioned, appears as a common feature within the literatures on normative and applied ethics. We must find a way to reassess a role for technology in shaping moral choices that is not present in a checklist approach. Secondly, the focus of this book is upon group deliberation and participatory-deliberative decision-support, rather than individual decision-making. Rather than trying to reach a specific conclusion on individual behaviour as I mentioned in chapter 3, the concern is with highlighting the realm of the ethical within a broader discussion of the socio-technical implications, governance and policy context of SECT. A key methodological difference between what is presented in this chapter and much of the ethical decision-making literature, is that deliberative decision-support must involve elements of iteration, facilitated dialogue between multiple participants and negotiation between competing values, judgements and principles. As I have established, deliberative methods allow for reflection, development and change in values amongst the participants. An iterative design would support this goal by allowing ideas to be expressed, evaluated and re-examined dialogically. Having shed the various forms of top-down applied ethics; the proposed reflective equilibrium-based model is a coherentist form of ethical reflection and deliberation. I suggest that this caters for the necessary iterative and reflective aspect, by allowing expressed moral judgements to be considered and amended or developed when appropriate within a discussion amongst deliberating actors.

5.4 The Structure of a Decision-Support Procedure

The first task is the formulation of a coherent theoretically informed structure. By explicitly adopting a pragmatist framework of ethical decision making, it is necessary to begin by establishing bottom-up problem framing by grounding deliberation on ethical issues within the practical techno-scientific and socio-political decision context. The first task is effective information provision to citizen participants in a participatory-deliberative process, and this involves a balanced range of information resources and opportunities to allow participants to assess their own information needs, question experts and to prepare for informed deliberative engagement. This phase is of critical importance, and one that theorists of deliberative democratic modes of governance often overlook. Deliberative processes involve both internal reflection and public discussion, though deliberative theorists and practitioners tend to focus primarily on the discursive component. Goodin and Niemeyer's (2003) study of citizens' juries on Australian environmental issues show how jurors' attitudes changed more in the context of the 'information' phase of the jury proceedings, involving a large degree of 'deliberation within', than during the formal 'discussion' phase. Given the relative power that the information provision aspect has on deliberative quality and the transformation of participant values, the balance of information in terms of its type (scientific, ethical, political), source (from NGO, campaigner organisations, print media, scientific publication) and content (quantitative, qualitative, peer-reviewed, opinion piece etc), is highly important. Ensuring balance is an art rather than science, and elements of iteration and participant self-evaluation of information needs is a necessary component of ensuring deliberative success.

Within the discussion phase of the process deliberation around such aspects of SECT should involve techniques to elicit and record a (long) list, not only of identified stakeholder actors – which might include, amongst others, affected local communities, politically and economically marginalised groups, governmental and non-governmental organisations; but also other non-human components, including technological artifacts, designs, non-human organisms, ecosystems, and built and natural environments. This is what Actor Network Theorists term generalised symmetry, whereby technological artefacts and other non-human elements should be described in the same terms as human agents (Latour 1993). Thus the unit of reference is the 'actant', to borrow the ANT terminology. The task is then to identify the relationships between these heterogeneous actants. Tools such as stakeholder mapping (SM) (McElroy and Mills 2000) have frequently been used successfully to draw out the interests of different civil society actors, identify conflicting and collaborating interests and assess their roles at different stages in a decision-making process, and these will be discussed in further detail later on. SM focuses, however, solely upon the human elements of organisational relationships. The model could prove useful though when adapted to ethical PTA, because such a method encourages deliberating participants to examine synergistic relationships between different groups. The goal is to adapt this type of method to include other non-conventional elements, potentially including future-generational and environmental interests, and the technologies themselves. Also such methods could be adapted to ethical deliberation in a relatively simple manner by framing the analysis and mapping processes in terms of how the behaviours and structures of one group can be both ethically motivated and ethically consequential to other groups. Due to the complexity of the stakeholder categorisations that result from SM, it may be necessary to then cluster the results into conceptually contiguous groups for simplification and further ethical deliberation. Although this process is comparatively time consuming and complex, it is meta-ethically preferable to the simplified, arbitrarily selected and monolithic categories of 'stakeholders' presented in tools like the ethical matrix.

The next task is the identification of suitable principles. I have argued that meta-ethical justification of selected principles must be consonant with bottom-up deliberation. I intend a principlist approach that is applied in a manner congruent with the perspective of Beauchamp and Childress. They articulate ethics as a dialectical relationship between ethical principles and concrete ethical problems, where the emergence of new ethical problems provokes a critical analysis and possible reformulation of existing ethical principles. Like a number of applied ethicists, they assert that understanding ethical theory (and by extension ethical principles) as having a dialectical relationship with human practices will lead to a reformulation of such theories and may provoke a modified view of actual ethical problems. I take forward Beauchamp and Childress's claim that the examination of ethical problems should be a process, not the application of rigid ethical principles (Beauchamp and Childress, 2001). In light of this, I suggested in chapter 3 that Rawls's concept of 'reflective equilibrium' provides a suitable basis for grounding the selection and justification of ethical principles in a process that is sensitive to competing, participant-led, bottom-up moral judgements.

In the reflective equilibrium-based approach a selection of principles grounded in theory-based perspectives (that have been developed within a community of expertise i.e. the top-down part) is deliberated upon in reference to the communicative, dialogic and reflective aspects of public and stakeholder formulated moral judgements (i.e. the bottom-up part). In this and the following chapter, I examine how such a reflective equilibrium model can be operationalised; in other words, reformulated as a set of practical tools through adaptation based upon qualitative and deliberative methods for clarifying individuals' moral judgements and values, followed by the elicitation of a long list of ethical principles in order to provide the evaluation criteria against which these judgements are to be critically revised. By applying the range of identified principles to the judgements elicited through group deliberation and subsequent reflection upon the context of the principles in relation to the judgements themselves (and the specificities of the case), the outputs would be a series of 'considered' judgements that are coherent with a set of participant-selected and adapted principles that are in turn, case-specific and relevant to the technology in question. By adopting this type of structure we can open-up ethical deliberation to creative and imaginative ethical reflection that is context specific and theoretically grounded.

I also propose that the 'outputs' of a reflective equilibrium-based deliberative process must then be formulated into a series of ethically informed policy options or alternatives, by reflecting upon the practical implications of their implementation. It is necessary to pragmatically re-contextualise the more abstract elements of ethical deliberation back within the political, social and techno-scientific context of decision-making in a manner congruent with philosophical pragmatism. I therefore also consider the use of valuation techniques to ascribe weight to different options identified through the deliberative process and hence 'close down' the discussion to either an agreement on ethically informed actions (resulting in a specific policy option), or else a narrowed range of policy options based upon ethical 'criteria' identified throughout. I propose that when such tools are used in concert, this provides a participatory ethical assessment of SECT which is both meta-ethically justified and compatible with the structure and processes of PTA.

5.5 Choosing the Right Deliberative Tools

Running an effective PTA process involves selecting the right methods to facilitate dialogue amongst participants. It is important to consider that method selection is something of an art rather than a science, and that finding the correct technique for a particular context is inevitably problematic. Technique selection must be in response to a specific situational, practical and theoretical context. This task is difficult, as I mentioned in chapter 2, because the motivations for engagement are complex and multi-faceted, ultimately dependent upon who is doing the implementing (Governmental, academic, community organisation or industry-led), the stage at which public actors are involved ('upstream' at the stage when technological programmes are being designed, or 'downstream' at the stage where they are introduced into society), and the degree of decisional influence that they have (are they simply being informed about developments, or are they being made partners in the process?). Defining this level of engagement is dependent upon various ethical, cultural and political influences from across a wide spectrum of interests including pressure groups, NGO's, governmental agencies, academia and local citizen groups, all of which have a stake in the decision outcome and have different expectations of involvement in any given circumstance. When designing an ethical tool based approach it is necessary to build in an element of flexibility in the design in order to maintain compatibility with a range of different decision-contexts and other forms of participatory-deliberative methods.

The problems of process design are exacerbated by the fact that there are no quantifiable means with which to select the right techniques to facilitate engagement and decision-making in any given situation. No single benchmark or metric for evaluating the effectiveness or usefulness of any specific deliberative method exists in the academic or policy literatures (Lowndes 1998; Rowe et al. 2005; Rowe and Frewer 2000). Consequently there is no one-size-fits-all technique that can be considered 'best' for use in all circumstances, nor is there an established toolbox of techniques that can be drawn upon. Selection of a suitable dialogue technique depends on the circumstances, the purpose of the process, and consequently the nature of the results expected or required. This is then dependent upon the numbers of people to be involved, the timescale of the process, the geographical spread of participants, the complexity of the issue, the involvement of specialists and the point in the policy process at which the engagement takes place. Deliberative methods can take place on any scale - from a dozen or so participants (e.g. a citizens' jury), hundreds (e,g, consensus conferences or deliberative polls), or thousands (such as citizens' summits, or deliberation days). A process may be a one-off event, or part of a series of activities running over several years. Each method has a specific design format involving different types of information provision, levels and types of knowledge, participant numbers and demographic characteristics. Many are designed for specific functions and the proprietary formats may not be translatable to ethics-specific deliberation. To keep things broadly generic and hence flexible in the face of these varying factors, I present an ethical evaluation process in the format of a deliberative workshop.

5.5.1 Deliberative Workshops

In its generic format a deliberative workshop denotes a qualitative approach that brings together a group of people usually for a single day. Workshops are collaborative processes where researchers and participants work intensively upon an issue or question. They combine elements of qualitative research, brainstorming and problem-solving; often involving larger numbers of participants than conventional focus-groups and using more than one moderator or facilitator. They allow time to explore the attitudes, values and beliefs of participants and also provide them with information and arguments in order to reach a critically informed position. As workshops often last longer than focus groups or interviews, this adds a level of flexibility because it is possible to vary the composition of the workshop depending upon the size of the participant groups, divide tasks throughout the day's deliberation and divide larger groups up where necessary. The longer time frame also allows moderators or facilitators to challenge the positions of participants as the day progresses, for example by introducing different types of information throughout the session, or by allowing time for presentations and plenary question-and-answer sessions.

Deliberative workshops have their roots in James Fishkin's (1995) work on 'Deliberative Opinion Polls' and more recently on Citizens' Juries (Fishkin et al. 2000; Smith and Wales 2000). Fishkin's work concerned the tendency of conventional opinion polls or focus groups to gauge 'what people think' about an issue, when they are responding essentially in an uniformed state. He sought to develop ways of allowing participants to not just state their preferences amongst a set of externally defined options, but to reflect on the core issues and creatively problem-solve to find suitable solutions. The work was instrumental in bringing deliberative methods into practical policy problems, and in showing how they provide both a richer picture of the participants' views and values towards an issue and can provide creative input to decision-making situations.

What distinguishes a workshop from a focus-group or group interview is that it involves a series of discussion activities, using different groupings, techniques and contexts, rather than simply 2-3 hour recorded small group discussions that often have no need for hands-on practical involvement, special materials or facilitators. This allows time to consider the details of an issue rather than encouraging participants to offer shallow, off-the-top-of-the-head reactions or beliefs in the way that attitude assessment methods such as surveys, opinion polls or focus groups might. In conventional attitude assessment methods, the response of participants is regarded as an indicator of something else - what they think, experience or do. This is a largely due to the theoretical legacy of behaviourism in sociology and social psychology, and is common in many psychological and social scientific research methods. Attitude assessments are often used as tools to gain access to some state of affairs which is deemed to exist independently of participants' verbal or textual representations of them; by contrast, deliberative workshops allow broader development of attitudes and values over a longer period of interactive dialogue. It also becomes possible to see whether and how these can change and what arguments and information have had the greatest impact. Crucially, deliberative workshops also provide a forum for participants to be challenged by one another, thus encouraging the development of ideas and beliefs. The advantage of the deliberative workshop design is that it allows analysis of the richer, highly interactive and iterative process by which participants (ethical) values are constructed through dialogue. The point to take away from this is that deliberative workshops allow the progression and development of constructed values through dialogue and reflection, rather than categorising them as simple statements of 'preference' that can be 'elicited' as a static snapshot of their innermost thoughts (Fischhoff 1999, 1991; Gregory et al. 1993). In short, this is what differentiates a deliberative process from conventional attitude assessment methods: values are perceived as malleable, rather than static positions that can be drawn out of people by asking the right questions.

In light of these advantages, the aim is to design a deliberative workshop that facilitates ethical reflection, collaborative discussion and critical decision-support that is inspired by reflective equilibrium – transposing the concepts of coherentist ethical justification from an individual practice to a framework to facilitate group deliberation amongst citizens.

5.6 Conclusion

In the following chapter I turn to the method and practice of operationalising reflective equilibrium into a set of deliberative tools. My aim is to develop an approach that sequentially fleshes out the issues by first establishing the inter-related socio-technical aspects of the SECT problem in question through group deliberation; secondly to draw out the ethical issues that relate to the problems identified, by eliciting contextually relevant moral judgements; thirdly to establish a coherent set of principles against which to evaluate the judgements, followed by a deliberative process framing these principled judgements as alternative strategies and then to weight and score them in an iterative process that highlights future policy options, recommendations and areas for future research. The general pattern follows a fourteen stage process:

1. Establish a participant-led dialogue process concerning the socio-technical issues of the socially and ethically contentious technology under consideration
2. Draw up a list of questions and ideas around which to formulate group discussion of ethical issues
3. Identify a range of actants: technologies, social actors, affected organisations and environments
4. Assess socio-economic, political and techno-scientific information
5. Discuss the implicit and explicit ethical issues, concerns and questions raised
6. Discuss individuals' judgements and intuitions on these issues
7. Scope a list of related moral principles, and amend the principles where appropriate

8. Apply principle to judgement and judgement to principle
9. Assess principled judgement and assess situated principle
10. Repeat steps 8 and 9 until an equilibrium is reached
11. Evaluate the coherent positions and their applicability as decision-making options in technology policy
12. Attribute weights to the options through group voting (such as nominal group technique)
13. Encourage feedback on the outcomes of the decision-making process.
14. Assess further areas of related ethical inquiry

The model displays aspects of the step-wise decision-making structure of the checklist-type approaches to ethical decision-making combined with some of the principlist features of the ethical tools mentioned in the previous chapter. Participants move through a sequential decision-process, beginning with a general discussion, identification of issues, affected actors and artefacts, the drawing out of implicit ethical issues, reflection on relevant principles and personal reflections in the form of moral judgements, followed by a weighting and decision procedure that reintegrates the ethical deliberation to practice by highlighting practical steps for technology governance based upon the preceding steps.

When operationalising the reflective equilibrium procedure to technology assessment, the emphasis is upon examining the relationship between methodologies to facilitate ethical reflection and the broader field of participatory-deliberative decision-making processes. In the following chapter I focus upon the development of multi-staged iterative evaluation and reflection upon the values and judgements of the participants and the moral principles involved in a way that can be applied to the practice of PTA. What makes this process unique as an approach to ethics is that it is done as a group-based deliberative procedure that combines elements of issue and stakeholder mapping, reflective group discussion, evaluation and decision-support. Therefore, in reference to the combination of elements from reflective equilibrium and group based participatory-deliberative methods, the framework for a toolkit approach call a reflective ethical mapping (REM) procedure. The following chapter focuses upon the discussion and development of suitable ethical deliberative decision-support tools that can fit in to this procedural ethical participatory technology assessment process, with examples drawn from empirical work around public reflections on the ethical issues surrounding long-term radioactive waste management in the UK.

References

Bowen, S.A.: A Practical Model for Ethical Decision Making in Issues Management and Public Relations. Journal of Public Relations Research 17(3), 191–216 (2005)

Burgess, J., Clark, J.: Evaluating public and stakeholder engagement strategies in environmental governance. In: Peirez, A.G., Vas, S.G., Tognetti, S. (eds.) Interfaces Between Science and Society. Greenleaf Press, London (2006)

Burgess, J., Stirling, A., Clark, J., Davies, G., Eames, M., Staley, K., Williamson, S.: Deliberative mapping: a novel analytic-deliberative methodology to support contested science-policy decisions. Public Understanding of Science 16(3), 299–322 (2007)

ERC. 2004. PLUS - A Process for Ethical Decision Making (2004), http://www.ethics.org/plus_model.htm (accessed December 01, 2004)

Fischhoff, B.: Value elicitation: Is there anything in there? American Psychologist 46, 835–847 (1991)

Fischhoff, B., Welch, N., Frederick, S.: Construal processes in preference elicitation. Journal of Risk and Uncertainty 19, 139–164 (1999)

Fishkin, J.: The Voice of the People. Yale University Press, New Haven (1995)

Fishkin, J.S., Luskin, R.C., Jowell, R.: Deliberative polling and public consultation. National Centre for Social Research 53(4) (2000)

Forester- Miler, H., Davis, T.: A Practitioner's Guide to Ethical Decision Making (1996) http://www.counseling.org/docs/ethics/practitioners_guide.pdf?sfvrsn=2 (accessed September 10, 2012)

Goodin, R.E., Niemeyer, S.J.: When Does Deliberation Begin? Internal Reflection versus Public Discussion in Deliberative Democracy. Political Studies 51(4), 627–649 (2003), doi:10.1111/j.0032-3217.2003.00450.x

Gregory, R., Lichtenstein, S., Slovic, P.: Valuing Environmental Resources: A Constructive Approach. Journal of Risk and Uncertainty 7, 177–197 (1993)

Jones, T.M.: Ethical Decision Making by Individuals in Organizations: An Issue-Contingent Model. The Academy of Management Review 16(2), 366–395 (1991)

Latour, B.: We have never been modern. Harvester Wheatsheaf, Hemel Hempstead (1993)

Lowndes, V., Stoker, G., Pratchett, D., Wilson, D., Leach, S., Wingfield, M.: Enhancing public participation in local government: A research report. Department of Environment, Transport and the Regions, London (1998)

Marshall, J.: An Ethical Decision-Making Model: Five Steps of Principled Reasoning (1999), http://www.ethicsscoreboard.com/rb_5step.html (accessed)

McElroy, B., Mills, C.: Managing stakeholders. In: Turner, R.J., Simister, S.J. (eds.) Gower Handbook of Project Management, pp. 757–777. Gower Publishing Limited (2000)

Nijkamp, P.: Multicriteria analysis: a decision support system for sustainable environmental management. In: Archibugi, F., Nijkamp, P. (eds.) Economy and Ecology: Towards Sustainable Development. Kluwer, London (1989)

Potter, R.B.: The origins and applications of "Potter Boxes". In: State of the World Forum. San Francisco, CA (1999)

Rowe, G., Frewer, L.J.: Public Participation Methods: A Framework for Evaluation. Science, Technology & Human Values 25(1), 3–29 (2000)

Rowe, G., Horlick-Jones, T., Walls, J., Pidgeon, N.: Difficulties in evaluating public engagement initiatives: reflections on an evaluation of the UK GM Nation? public debate about transgenic crops. Public Understanding of Science 14, 331–352 (2005)

Smith, G., Wales, C.: Citizens' Juries and Deliberative Democracy. Political Studies Review 48(1), 51–65 (2000)

Stirling, A., Mayer, S.: A Novel Approach to the Appraisal of Technological Risk: a Multi-Criteria Mapping Study of a Genetically Modified Crop. Environment and Planning C: Government and Policy 19(4), 529–555 (2001)

Thomson, A.: Critical Reasoning in Ethics: a practical introduction. Routledge, London (1999)

Van-Hoose, W.H.: Ethics and counseling. Counseling & Human Development 13(1), 1–12 (1980)

Chapter 6
Opening Up Ethical Dialogue

6.1 Introduction

In the previous chapter I outlined the decision framework for a reflective ethical mapping (REM) procedure based upon the Rawlsian concept of reflective equilibrium. The following two chapters 'operationalise' this decision framework by outlining a series of practical deliberative methods that can structure and facilitate this type of coherentist ethical reflection in a group setting. Each of the methods presented in these chapters is proposed for the context of a deliberative workshop – a series of small group discussion activities run with a small number of participants over one or two days. The choice of participants is of course context specific, and these methods can potentially be adapted for both expert and non-expert use. The methods proposed are, however, principally designed with the non-expert public stakeholder in mind. I have argued that this group of stakeholders must be engaged with on these issues in order to avoid the technocratic decision-making based solely upon the voice of experts (in this case perhaps philosophers rather than engineers or scientists), and to ensure strong deliberative democratic control of socially and ethically contentious technologies (SECT).

When operationalising reflective equilibrium as a decision framework the outline procedure has four basic stages:

1. To identify relevant topics and issues for discussion, highlight relevant 'actants' – the stakeholders, natural and social systems and technological artefacts involved, to produce an actor-network of cause and effect relationships and resultant ethical issues.
2. To 'elicit' or stimulate participants to consider their moral positions and to make specific moral judgements about the technologies in question, the motivations of different actors and the 'scripts' of technological artefacts.
3. To relate these moral judgements to a series of ethical principles grounded in common sense ethics, broadly representative of different dominant theoretical traditions in normative ethics, and to iteratively discuss the implications of these principles to the judgements elicited in the previous stage. Then to recontextualise or otherwise amend these principles in light of new insights drawn from the discussion of specific political decision-making cases and specific technologies.
4. To identify strategies and options for political decision-making that are case sensitive, grounded in the consideration of principles and personal

M. Cotton, *Ethics and Technology Assessment: A Participatory Approach*,
Studies in Applied Philosophy, Epistemology and Rational Ethics 13,
DOI: 10.1007/978-3-642-45088-4_6, © Springer-Verlag Berlin Heidelberg 2014

moral judgements; to use weighting and option appraisal mechanisms to decide between different courses of action. To consensually agree within the group as to the course of action available, and then to reflect upon these courses of action and potential issues that may arise in the future and hence where further deliberative ethical reflection is necessary.

In the following two chapters I present and discuss a series of deliberative methods that can fulfil the criteria for this sort of decision-making structure. This methodological discussion is divided into two sections. Stages 1 and 2 are discussed here in chapter 7 in the context of opening up deliberation; in the sense of describing methods that can elicit new information, reveal new conceptual categories, options and ideas, and thus illustrate the sociotechnical complexity of the SECT in question. In chapter 8, stages 3 and 4 are discussed in the context of closing down the discussion, in the sense of describing methods which evaluate concepts, identify and weight options and alternatives and reach (tentative) conclusions.

It must be noted that in the spirit of pragmatic philosophical inquiry it is not intended that these methods be considered definitive, nor complete. It represents an experimental model of ethical decision-making which can be adapted, added to or amended with further exploration, testing and context-specific reflection. The methods presented here are therefore presented as template for an ethical toolbox, where the tools can be refined, expanded or removed for different sorts of decision tasks. Also in the spirit of pragmatic inquiry, these methods are applied in context; throughout the discussion, examples drawn from empirical data collected from three deliberative citizen-stakeholder workshops are used to illustrate the Reflective Ethical Mapping (REM) procedure in practice. The case study under consideration concerns the long-term management of long-lived radioactive wastes arising from nuclear power production in the UK; and the following section presents something of a preamble – outlining the political and ethical context in which radioactive waste management decisions have been taken.

6.2 Ethics, Technology and Environment – The Case of Radioactive Waste Management in the UK

The long-term management of the United Kingdom's legacy of radioactive wastes is a controversial environmental management and technology governance issue. UK radioactive wastes result from the production of nuclear energy, the manufacture of nuclear fuels, spent fuel reprocessing, industrial applications, military activities, research and medicine. Radioactive wastes contain materials that are atomically unstable and release ionising radiation that has the potential to damage DNA, which in acute doses can cause radiation sickness and other ill-health effects, and over the longer term can increase the risk of malignant cancers in those exposed to significant doses. These hazardous end-of-pipe pollutants generated primarily from activities associated with nuclear power have therefore created significant problems for political administrations in the UK and for other nuclear producing nations throughout the developed world. To date, wastes are stored at 34 locations around the UK, awaiting the construction of a long-term radioactive waste management (hereafter referred to as RWM) facility. However, the implementation of a long-term technological strategy and site selection process for RWM facilities (referred to as 'siting') has remained a significant environmental and political challenge, with no agreed site for a facility yet decided.

Though this issue has long been a source of political gridlock (see for example Kemp et al. 1986; Atherton and Poole 2001; Blowers and Pepper 1988; Blowers 2010), considerable progress towards implementing a solution has been made in recent years. The UK Government's "Managing Radioactive Waste Safely" (MRWS) programme (DEFRA 2001, 2007; CoRWM 2006; BGS 2010) is an initiative seeking to establish a socially legitimate and technologically sound long-term solution. The MRWS programme is something of a departure from previous RWM policy strategies implemented since the late 1970s. Historically, UK RWM policy has been approached from a primarily scientific and technical standpoint. Radioactive wastes are produced primarily through industrial processes and thus have often been treated as a technical problem. The primary role of RWMOs has typically involved research into disposal techniques followed by siting processes aimed at finding suitable locations for wastes based primarily on outcomes that presented the lowest potential 'risk' according to the best available scientific evidence and technical criteria. Such an approach has been frequently criticised as being technocratic, because it fails to address significant concerns amongst communities affected by siting in their local area, alongside broader societal concerns about how best to manage these wastes whilst maintaining long-term public safety (these issues have been extensively discussed by Petersen 2001; Dunlap 1993; Peelle 1987; O'Hare et al. 1983; Blowers et al. 1991; Blowers and Sundqvist 2010). With local conflict over technocratic siting proposals for RWM facilities repeatedly blocking attempts to identify suitable sites for what are termed low and intermediate level waste disposal[1], Government adapted its approach and

---

[1] Radioactive wastes are classified according to the levels of radioactivity that are produced (Nirex 2002):

- **High Level Waste (HLW)** – Radioactive wastes in which the temperature may increase significantly as a results of radioactivity. Liquid High Level Waste can be in the form of nitric acid solutions containing fission products created by reprocessing irradiated nuclear fuel.
- **Intermediate Level Waste (ILW)** – Has lower levels of radioactivity than the HLW and significant heat is not a factor in storage and disposal. This includes a variety of wastes such as chemical sludges, metals (mainly in the form of fuel cladding, fuel element debris, plant items and equipment), and graphite from reactor cores.
- **Low Level Waste (LLW)**- The major components of LLW are soil, metals and building materials. Low Level Wastes consist of those that are unsuitable for disposal with ordinary refuse, but within technical specification do not exceed 4 GBq (giga-becquerels) per tonne of alpha, or 12 GBq per tonne of beta/gamma activity
- **Very Low Level Waste (VLLW)** - Wastes that can be disposed of with ordinary refuse, each 0.1 cubic metre of material containing less than 400 kBq (kilo-becquerels of beta / gamma activity) or single items containing less than 40 kBq.

radioactive waste management organisations (RWMOs) such as the former UK Nirex Ltd. subsequently sought to reframe the problem as a 'socio-technical' policy issue (Flüeler 2006; Flüeler and Scholz 2004), opening up RWM policy-making to a broader range of actors and viewpoints (Lidskog 1997; Gunderson 1999; Litemanen 1996; Freudenberg 2004; Atherton and Poole 2001), and shifting emphasis towards incorporating political, psychological, social and ethical factors (Sjöberg 2003; Atherton and Poole 2001; Carter 1989; Kemp 1992; Slovic et al. 2000). This has been realised in practical terms through an implicit political commitment to sustained and inclusive public and stakeholder engagement (PSE) on social and ethical issues and the incorporation of diverse values and viewpoints into decision-making processes (Gemmell 2005; Chilvers et al. 2003; Flüeler 2005; Sundqvist 2005; CoRWM 2005; Burgess et al. 2004; Cotton 2009). Consequently, there has been a significant trend towards the use of PTA methods designed to facilitate the integration of community and stakeholder values into decision-making processes. Concerns over the health risk implications of radioactive wastes are also linked to questions of social legitimacy and procedural fairness in relation to who gets a say in how radioactive wastes are managed (Andrén 2012).

These justice concerns regard the physical attributes of radionuclides in the natural environment, but also the influence of RWM facilities on the values, perceptions, place attachments and judgements of the citizens exposed, as communities can often become stigmatised by facilities sited in their locality (Gregory and Satterfield 2002). Thus, RWM policy decisions are fundamentally ethical in character, and explicit ethical justification within the political decision-making process is required.

This notion of ethical legitimacy in the technology management processes associated with radioactive wastes has been recognised by both national and international authorities. Notably the OECD Nuclear Energy Agency (NEA) and the International Atomic Energy Agency (IAEA) defined three groups of ethical issues related to RWM which have informed policy development in nuclear power producing countries (NEA 1995; IAEA 2002):

- Intra-generational equity – defined as proper access to the decision-making process for all stakeholders; transparency and accountability on the part of the relevant authorities when taking those decisions; a fair distribution of the disadvantages of activities such as those that produce radioactive wastes; the 'polluter pays principle'; compensation for affected communities.
- Inter-generational equity – defined as protecting the interests of future generations who have not (or may not be) benefited from the deployment of civil nuclear energy but may have to deal with the legacy.
- Environmental equity – a growing (though perhaps still not firmly established as yet) belief that concern should be paid not only to the welfare of human beings now and in the future but also to other living species and to the environment in a wider sense.

These so-called 'ethical principles' (as the IAEA defines them) were later adopted by the Government-appointed Committee on Radioactive Waste Management (Blowers 2006; Grimstone 2004) referred to as CoRWM (pronounced 'corum') – an independently facilitated expert committee charged with assessing the options for radioactive waste management (including deep geological disposal in an underground facility, disposal in ice sheets, in space, or subduction zones between tectonic plates). CoRWM used these principles as categories of ethical issues to be explored in the RWM options assessment process, alongside their work on engaging public and stakeholders in the examination of potential technological options.

6.2.1 CoRWM's Work on Ethical Issues

CoRWM recognised that ethical considerations would inevitably have an important part to play in its decision making process, and so they formed a key component of a set of Guiding Principles that informed the structure of their work (Grimstone 2004; CoRWM 2004). The Guiding Principles were described as statements of fundamental core values (Blowers 2006), and applied very broadly to CoRWM's working practices, intentions and their approach to the PSE process (Blowers 2006; CoRWM 2004):

- To be open and transparent
- To uphold the public interest by taking full account of public and stakeholder views in our decision making
- To achieve fairness with respect to procedures, communities and future generations
- To aim for a safe and sustainable environment both now and in the future
- To ensure an efficient, cost-effective and conclusive process.

At the heart of these guiding principles were an underlying set of ethical values, specifically codified as working practices. However, these principles are provide codes of conduct rather than tools for the assessment of ethical criteria in relation to decision-making over which technological option to choose. It was important for CoRWM to clearly state the principles that underpinned their procedures. However, these principles alone were insufficient for assessing the wide ranging issues involved in participatory technology assessment. Thus part of CoRWM undertook specific work in this area of technological and environmental ethics.

During CoRWM's PSE programme, the ethical concerns associated with RWM options were identified. The criteria used for short-listing options therefore specifically incorporated ethical aspects from the start. CoRWM began by gathering feedback from PSE events involving roundtables, open meetings, citizens' panels and the national stakeholder forum, as well as a wide range of written and website responses (Blowers 2006). Also, ethical discussions of the option assessment specialist panels took place on a range of topic areas (including safety, transport, site security, environmental and socio-economic impacts, implementability etc.) and these were a key aspect of the multi-criteria decision analysis (MCDA- itself a form of Technology Assessment) process undertaken for choosing amongst the different options available.

CoRWM's programme of specialist ethics and social science input was linked most directly to a stage they termed 'Holistic Analysis', that broadly took account of combined technical knowledge, PSE input and CoRWM members' views on a range of issues such as storage lifetimes, the extent to which institutional control over a facility could be guaranteed into the future and the option to retrieve the waste from an underground facility (CoRWM 2006). They used MCDA to address ethical issues directly, and through the weighting of outputs, the implementation recommendations (which involved interim surface storage of radioactive wastes followed by long-term deep geological disposal) drew heavily on ethical input (Collier 2006; Blowers 2006).

In September 2005 CoRWM held an external ethics workshop, and this was to be the main vehicle for specialist input on ethical issues (ibid). It brought together Members of CoRWM and various UK and international specialists in the ethical issues associated with radioactive waste management (including philosophers and sociologists). The overall aim was to "explore the ethical aspects of radioactive waste and in doing so to (Blowers 2006):

- Help [Members] understand the importance of ethical considerations and how they may be taken into account;
- Inform and generate discussion on ethical issues to enable CoRWM, stakeholders and the public, to think about the ethical aspects of the different options for managing radioactive waste, and thereby;
- Provide an input into the PSE round associated with options assessment and to reflect on outputs from earlier rounds of PSE;
- Understand how ethics need to be integrated with scientific outputs in a process of holistic decision-making".

This workshop involved firstly developing a 'briefing pack' of CoRWM and participants' perspectives. The workshop itself took the format of a series of presentations and discussions on four main topics (Blowers 2006):

- In what ways is radioactive waste an ethical issue?
- Inter-generational equity
- Intra-generational equity
- Ethics and environment

After a process of deliberation, external participants were also asked at the end for their intuitive preference amongst the short-listed options. Following the workshop, a report was made along with a video that was subsequently shown to a series of Citizens' Panels (Collier 2006). This initial workshop was then followed by two option assessment 'ethics sessions'. At a plenary session in 2005 CoRWM Members considered the pros and cons of the short-listed options against a set of ethical tests based on the concepts surfaced at the workshop. The plenary then considered the options against a set of environmental principles based in part upon the workshop outputs. As a result of the specialist input to the options assessment process and the feedback from the PSE programme, these events (and the feedback that followed) led CoRWM to conclude that, "all in all, the ethical dimension of decision-making has played an integral role in the CoRWM process" (Blowers 2006).

In many respects, the ethics programme that CoRWM implemented was a success. Input from the public through the PSE phases and then specialist input from experts was incorporated into the decision-making process. As a result, ethics became a serious criterion for the technology assessment of different management options, and questions over aspects of intergenerational equity became a primary discriminating factor between the choice of final deep geological disposal of wastes and a long-term storage solution (Blowers 2006). Thus it could be argued that the ethical components of sociotechnical radioactive waste management design were assessed. However, in this respect the CoRWM ethics evaluation process is an illustrative example of the limitations of top-down ethics discussed in chapter 3, as the assessment process was based upon the input of specifically identified ethics experts. In CoRWM's programme there was an early stage of public and stakeholder involvement on the ethical issues in the PSE programme; when defining the broad area of work and issues to be examined. When it came to examining specific ethical issues in greater detail for their Holistic Assessment and MCDA stages, CoRWM chose to base its ethical evaluations primarily on the advice of specialists rather than that of citizens (Cotton 2009). Adopting a similar approach towards the issue of implementing a long-term RWM strategy (at the stage of site selection) would, however, likely become fraught with both philosophical and political difficulties. As shown in chapters 2 and 3, basing decisions about RWM technology strategy and facility siting primarily upon technical expertise will likely lead to the rejection of siting proposals and to a political backlash against the RWMOs involved, as has been seen in all previous examples of radioactive waste siting in the UK, and in other developed nation contexts (Blowers and Pepper 1988; McCutcheon 2002; Kemp 1992). If the technical expertise under consideration is ethics-based rather than science-based, one could surmise that a similar process of local backlash would occur, with communities objecting to the notion that an outside body could decide not only what is safe, but also what is fair for the community in question. Thus, the case of radioactive waste management is illustrative of a need for bottom-up community and stakeholder engagement for ethical evaluation as part of a PTA process.

6.3   The Empirical Context – Examining the Ethics of Radioactive Waste Management in Nuclear Communities

In each section of the subsequent methods discussion I present some of the findings emerging from three day-long workshops held in communities in close proximity to nuclear power stations[2]. Locations close to existing nuclear power facilities were selected based upon the assumption that such so-called 'nuclear

---

[2] **Workshop details:**
The workshops were held in the communities of Leiston, Aldeburgh (both in proximity to the site of the Sizewell nuclear power station in Suffolk, southeast England) and Hartlepool (home to nuclear power station currently being decommissioned and a neighbouring community to the town of Billingham, a previously proposed ILW facility site in the northeast of England).

Aldeburgh and Leiston workshops
The first community workshops were held in the Suffolk coastal town of Aldeburgh on 3rd February 2007 and in neighbouring Leiston on 10th February 2007.

Participants: The first and second workshops ran with 10 participants, an even split 5 male: 5 female, with ages ranging from 28-84. There were 11 participants recruited in total, 9 of which attended both sessions and 2 attended one session each (one on the 3rd and one on the 10th). In short informal interviews with participants prior to the workshop, one participant declared a strong 'pro-nuclear' stance, and two others a strong 'anti-nuclear' stance, with no other participants expressing such viewpoints. The sampling of participants was based upon attaining a broad range of perspectives on the issues, at no point was the workshop intended to be demographically or statistically representative and this fact was made clear to participants upon recruitment. Each participant was paid £110 for their participation in both workshops (the two that attended one workshop each were paid £55).

Location: Participants were recruited from Aldeburgh, Leiston and Thorpeness in Suffolk. Both communities are within a 5 mile radius of the Sizewell nuclear power station. Given the history of local nuclear power generation and that the power station was undergoing consultations on decommissioning throughout the research period, local nuclear issues were being discussed in stakeholder engagement forums, the local media and highlighted through protest actions (by the Shutdown Sizewell campaign for example). All of this contributed to a local 'buzz' about nuclear site management.

The first workshop took place at the community hall adjacent to the Church of St. Peter and St. Paul in Aldeburgh, a town situated 3-4 miles away from the Sizewell power station. The second was held at the Fairfield community centre in Leiston, approximately 2 miles from Sizewell power station.

Focus Participants were informed that the workshop would be running over two weeks with a slightly different topic focus in each session. In the first session the focus was upon national-level RWM implementation; specifically the ethical issues around site selection, the decision-making process and any issues that would apply to the UK as a whole. The second session focussed upon local-level issues, framed by the hypothetical question, "what would happen if waste management facilities were to be constructed in the local area?"

Hartlepool workshops
Participants: The final workshop ran with 8 participants, an even split 4 male/4 female, ages ranging from 32-88. Participants were paid £80 for attendance at 1 workshop (an increase on the previous two workshops, in order to gain greater attendance following prior difficulties with recruitment). The smaller group size was based upon two factors, firstly an 8-person group was more easily managed by a single facilitator, and secondly it alleviated financial constraints due to increased participant fees.

Location: The workshop took place at the Hartlepool Historic Quay, approximately 3 miles from Hartlepool power station.

communities' could represent suitable proxies for future radioactive waste facility hosts. The site communities were locations where existing nuclear facilities were being (or were soon to be) decommissioned. It was hoped that existing engagement processes around decommissioning (including site-use consultations), would help to generate interest in the workshops and encourage participation by local community members. Local changes in land use, employment patterns, property values and regeneration strategies related to nuclear development were likely to be discussed in local political forums and the local media in the selected locations. As RWM is an important facet of the decommissioning process, it was assumed that RWM would be perceived as a salient issue for these nuclear communities. By providing a forum for participants to discuss their concerns, values and judgements, it was again assumed that this would be a suitable motivating factor to improve participant 'recruitment' in those areas (in addition of course to the small cash incentives offered).

6.4 Engagement Methods

6.4.1 Actant and Issue Mapping

When beginning to assess the ethical issues involved in the management of SECT it is necessary to begin by trying to understand who and what is involved in the development and governance of the technology itself. As previously mentioned, the concern here is upon understanding technology not solely as an asocial and amoral artefact, but rather with understanding it as socio-technical process, the features of which can be drawn out by paying attention to what STS scholars term an actor-network. The epistemology and methodology of Actor-Network Theory contain both material and semiotic components, that is, they are concerned with the co-constitutive relations between physical objects and concepts (Law and Hassard 1999). For example, nuclear power involves relationships between

---

Focus: The workshop, like the second Leiston workshop, focussed upon a hypothetical scenario involving local radioactive waste facility siting.

Hartlepool was selected for the following four reasons:

1. Hartlepool's proximity to Billingham (approx. 3-5 miles), a former potential RWM facility site in the 1980's.
2. It is a densely populated area, widening the scope for participant recruitment
3. The geographic and socio-economic character of the Hartlepool area (i.e. a post-industrial town) contrasts with the comparatively affluent and rural Suffolk coastal region.
4. A suitable recruiter was found at an affordable price in the local area, thus reducing the time constraints to the researcher working alone

As I lived in a neighbouring community for 20+ years, it was felt that a degree of knowledge about local issues would help to build common ground with the participants, especially given that many had no knowledge of the University of East Anglia, the host institution from where the research was based.

politicians, technical specialists, Geiger counters, mathematical equations, computer models, technical reports, economically marginalised communities, radioactive isotopes, and so on. The breadth and depth of these human and nonhuman relations constitutes an actor-network. I posit that understanding the nature of this network, even on a relatively shallow level can be beneficial to the deliberative process of ethical reflection because it contextualises technology as something inherently conceptual, value-laden and culturally situated, as well as material and technical. By encouraging citizens to reflect upon this co-constitutive relationship, we provide a suitable platform to question the governance and control of technologies as a process that requires ethical deliberation.

In practical terms, the first aim within a reflective ethical mapping process is to encourage participants to map out these material and conceptual relationships by identifying a range of stakeholders, environments, material conditions, technological artefacts and other related 'actants', based upon the previously mentioned position of generalised symmetry in explaining actor-artefact relationships (Latour 1993). Once this series of actants is identified it is important to map out the interconnective relationships between them in order to produce an actor-network map that presents these relationships in terms that are conducive to ethical evaluation. The goal of the first stage is not to explicitly talk about ethics *per se*, but rather to 'warm up' the discussion in a manner that will facilitate ethical reflection in the subsequent stages. The reason for exclusion of explicitly ethical criteria at the start is based upon pilot testing of the methods presented in this chapter. It was clear from participant feedback on the development of these tools that non-specialist participants are not comfortable or willing to begin from discussion of what are broadly perceived as abstract philosophical concepts and arguments. The process presented here, thus begins by opening up discussion through the examination of concrete problems and specific issues that emerge through deliberative dialogue. Thus the deliberation is grounded in an examination of real world socio-technical systems in a pragmatic manner. It is necessary, therefore, to begin by talking about the issues that they find important, map out the related actants, discuss the socio-technical issues and then to reflect, at the end of the first stage, what the ethical issues might be. In the following section I discuss a series of methodological tools that could be adapted to meet such demands.

6.4.2 Mapping Tools

The theoretical basis for the first tool in this process involves attention to three different, though conceptually related 'mapping' approaches. The focus on mapping implies a process of identifying not only a list of relevant concepts, but also the linkages between them:

- Stakeholder mapping
- Concept mapping
- 'Hexagons for system thinking'

Of these mapping approaches, the first, called stakeholder mapping, emerged in the organisational studies literatures to describe techniques for identifying and assessing the effects groups with different and often competing interests have on a company or other organisation. In particular these methods focus upon the power that specific interest groups can exert, the relative likelihood of each to use that power influencing organisational outcomes and the level of interest that they hold in the outcomes of particular decisions. These groups often include consumer organisations, NGOs, suppliers or community representatives. The goal of stakeholder mapping is gauge which individuals or groups of stakeholders hold the most power to influence the actions of the organisation, and thus allow the organisation to assess which stakeholders would need particular focussed attention.

Various models of stakeholder mapping have emerged, principally in differing diagrammatic forms. What each share is an identification of different groups, and the arrangement of these groups to show their influence either on a central organisation or else to show the synergistic relationships between different interested parties. The former tend to be represented either in a matrix style dividing stakeholder groups according to their level of interest and level or influence, or else in hierarchies or 'onion rings' (Alexander 2005) that show the most influential stakeholders near the top or centre (as per the method). In the latter there is a tendency to show stakeholder relationships as influence diagrams or webs (Coakes and Elliman 1999), where the relationships between them can be lain bare. As the focus within this first stage of the reflective ethical mapping process is to identify relational rather than power influences, it is this latter approach that is adopted here.

Stakeholder influence mapping tools share conceptual similarities to the second approach on the list, termed concept mapping. Concept mapping is a diagrammatic technique for organising and communicating the relationships between concepts, theories and ideas. It has developed in the field of educational studies as a way to increase meaningful learning of academic science, building upon the constructivist approaches of learning theory. Concept mapping is based on the idea that, in a learning context, individuals use their prior knowledge as a framework for understanding and incorporating new knowledge. Thus, meaningful learning involves the assimilation of new concepts and propositions into existing cognitive structures (Novak and Gowin 1996; Anderson et al. 2004; Ausubel et al. 1978). Conceptual mapping is used primarily to stimulate the generation of ideas and encourage creative input. It is often used as an exploratory tool for brain-storming, note-taking, knowledge creation (i.e. transforming tacit knowledge of participants into an organisational resource), mapping the knowledge of groups, or in communicating complex ideas (Novak and Gowin 1996). The mapping process involves generating and recording concepts, enclosing them in circles or boxes, connecting concepts with a line or arrow, linking two or more boxes together. Linking words or phrases specify the relationship between the two concepts, whereby an individual 'concept' is "…a perceived regularity in events or objects, or records of events or objects, designated by a label" (Novak and Cañas 2006). The label for most concepts is a word, although sometimes symbols such as + or %, or more than one word is used. 'Propositions' can be defined as, "…statements about some object or event in the universe, either naturally occurring or constructed. Propositions contain two or more concepts connected using linking words or phrases to form a meaningful statement. Sometimes these are called semantic units, or units of meaning." (Novak and Cañas 2006).

Concept mapping is a useful tool for group deliberative procedures as it outlines issues, shows clarity in the inter-relationships between them and simplifies communication of the identified relationships to an outside audience. One potential problem, however, is that concepts are inter-related in a static way. If a concept is written down and then joined by a linking label it becomes fixed. Rigidity is not conducive to the iterative development and reflexive change that strengthens deliberative dialogue. Hence, a moveable concept map is preferable to a drawn (and hence fixed) concept map. In light of this, I move on to consider a similar mapping tool called 'hexagons for systems thinking', which overcomes this limitation.

6.4.3 'Flexible' Concept Mapping

Dialogical processes amongst groups of participants tend to occur in an unstructured and linear fashion. In other words, conversations tend to move freely from one topic to another without a predefined focus or central concept to bring the disparate aspects together. In some qualitative and deliberative methods (focus groups for example) this can be both a benefit and a hindrance. As a deliberative process develops it increases in complexity as more information, values and concepts are brought in to bear on the issue under consideration. Linear dialogue is 'free', unhindered by external framing effects from researchers or other bodies implementing such activities. However, in a participatory-deliberative decision-making context, unstructured and unrecorded linear dialogue places an excessive burden upon the memory of the participants to recall the different topic strands and to hold these together in the mind (Kaner et al. 2007). Some dialogical processes rely heavily upon the memory skills of participants to maintain discussion focus, or else upon those of the facilitators to guide discussions. Psychological research has shown that when tracking conversations memory alone is an inefficient medium when unsupported by visual representation (Miller 1956; Avons and Phillips 1987; Kikuchi 1987; Phillips and Christie 1977). To counter the limitations of individuals' attention span and memory, group discussion statements in meetings, workshops or focus groups tend to be recorded on paper flip charts, white-boards, or computers. Visual recording improves the recoverability of the conversation, allowing third parties to review the outcomes of discussion. Hodgson (1992) suggests, however, that although the outputs are recoverable, the generation of a checklist or diagram produces inflexible results, in the sense that utterances, ideas and concepts become fixed, either in a particular shape (such as diagrams) or in a particular order (such as in a list). Conceptual mapping, mind-mapping and other similar group brainstorming techniques share this problem of rigidity.

Hodgson's answer to this problem of inflexibility is what he terms 'Hexagon Modelling'. As a method, it shares similarities with Buzan's (2003) Mindmapping or De Bono's (1985) Lateral Thinking techniques; and like these it has tended to be used largely in business and managerial contexts for brainstorming, strategy development and planning. Where the hexagon method differs is in style of implementation: using a series movable hexagons for capturing ideas - a flexible, visual medium for handling conversation content, as opposed to the static pencil-and-paper conceptual modelling techniques.

The hexagon method involves asking participants in meetings or deliberative discussions to write out on separate hexagonal shaped cards a series of key system concepts that relate to the problem situations under consideration. These hexagons can be simple post-it style sticky notes, or more sophisticated magnetic rewritable plastic hexagons or computer models. Either way, the participant is then instructed to group these hexagons into semantically contiguous groups, and to provide these groups with a category label. Once these clustered groups have been formed (using in Hodgson's model a maximum of 15 hexagons), then the participants are asked to draw links, i.e. arrows, between the hexagons or clusters that denote the most important relationships, causal or otherwise, between concepts. In practice the hexagon modelling process is divided into stages:

1. The initial phase involves recording individual ideas and potential solutions onto separate hexagons.
2. The hexagons are then clustered into groups around specific issues and then labelled in groups called "issue maps".
3. The issue map is used to create an 'influence map' whereby the relationship between two issues is detailed on a third linking hexagon. Where 2 hexagons are touching, there is a question of what would fit into the interconnecting space, also touching these two hexagons.

In a hexagon model, different colours represent different types of thinking in the problem-solving process shown in Table 6.1 (Hodgson 1992; Hodgson 1994):

Table 6.1 Hexagon mapping categories

Yellow	Lateral thinking	Opportunity spotting
Green	Imaginative thinking	Innovation
Purple	Strategic thinking	Directing
Red	Decision thinking	Action

Hare et al (2002) suggest that the hexagon method can quickly elicit ontological, relational, and general structural knowledge about contrasting systems from groups or individuals, and incorporate it directly into a graphical model for further discussion. It provides an engaging (and colourful) visual memory aid and a means to assess the mental models, i.e. the symbolic representations and explanations of individual participants' thought processes in understanding an external reality that users draw upon in their discussion. Hodgson's approach involves combining the three aspects of interactive and mobile representation, "effective thinking frameworks as transitional objects" and interactive facilitation skills (Hodgson 1994). In short, the advantages of the hexagon modelling technique lie in the straightforward approach, use of systemic conceptual modelling, colour-coded representation of concepts and flexibility of the hexagons concept models throughout the development of the discussion.

6.4.4 *Adapting the Hexagon Method for the Consideration of Ethical Issues*

Hodgson's hexagon model format is unsuitable in its current format for ethical deliberation. It is structured around a series of related concepts and aims to identify the different kinds of thinking (strategic, lateral etc...) involved and conceptually map them together, drawing inference between linked concepts with no explicit reference to ethics. The main strength of the system therefore lies in its visual representation rather than conceptual content.

In adapting the hexagon method, I present a model concerned with mapping out the relationships between the socio-technical issues identified by participants, followed by a problem identification or 'brainstorming' exercise identifying the various actants, a subsequent discussion of the interactions of cause and consequence that emerge, followed by the identification of specific ethical issues which can be carried forward for further discussion. The adapted ethical hexagon method has four main objectives, taken from the assessment of ethical tools and conceptual mapping processes. It is designed to incorporate the combined strengths of ANT analysis showing the relationships and 'actantiality' of both human and non-human components of actor-networks; and conceptual mapping techniques by showing the complex interrelationships between groups of diverse ideas and concepts and representing them diagrammatically; the hexagon modelling approach that allows flexibility and 'maneuverability' of concepts throughout a process of deliberation; the ethical matrix's structured approach showing the interrelationships between ethical concepts and stakeholders; and the discursive flexibility and colour-coded ethical concepts used by the ethical grid.

By combining these aspects, the idea was to develop the conceptual mapping approaches to specifically accommodate ethical reflection and discussion and also to break free of the confining grid (or matrix)-like structures of the aforementioned ethical tools. In this way, the first method uses a series of colour coded hexagon-shaped writeable sticky-backed notes or magnetic hexagonal discs. Each of the hexagons are given a colour category, for example in the format shown in Table 6.2.

Table 6.2 The structure of the hexagon method

1.	Issue identification	Map out the issues, questions and concerns of the participants in relation to the SECT in question	Blue
2.	Actants	Identify the actors, objects, beings, environments etc. that are affected by the management and implementation of the SECT	Yellow
3.	Actions, behaviours, intentions and procedures	Identify the possible intentions of actors, and the 'scripts' of technological artefacts: showing stakeholder relationships, motivations, procedures and rules, and influences on other actants within the network	Pink
4.	Consequences, outcomes or effects	Identify the potential consequences and outcomes of the actant relationships, and how this leads to further actions	Orange
5.	Ethical question and issue identification	Suggest the ethical implications of stakeholder cause and consequence interactions and raise points for further discussion	Green

The first category of hexagons represent the issues under consideration. This problem identification phase lays the groundwork for subsequent deliberation, providing a bottom-up framing of the problems from a citizen perspective. It must be stated that this bottom-up framing is in relation to information provided to the participants before the deliberation begins. By providing information materials, access to expert testimony and opportunities to question this testimony (in the manner of a citizens' jury), participants can form opinions on the issue that are deeply considered, rather than the shallow attitude assessments of focus groups. At the beginning workshop stage, the participants have a short discussion and are then handed a small stack of blue hexagons. They each individually write down the issues that they think are important to the discussion, identified from their own research and response to expert input (where available). These are then collected by a facilitator and clustered together on a board or blank wall. This clustering process is negotiated between the facilitator and the group to identify common themes and contiguous groupings of issues. Duplicates are removed or added to the cluster, and these clusters are then voted on using a system similar to nominal group technique (Delbecq and VandeVen 1971). Each participant is given a number of sticky dots to represent votes. The clustered issue groups are numbered and each participant silently gives each category of issues a number of dots. They can place the whole of their stake onto a single issue, or spread it between a number of issues. The voting process is reflective of the issue salience for each participant. Once this voting process is complete, the results show which categories of issues are transferred to another board for further discussion. The advantages of this process are twofold. Firstly, the individual identification of issues and silent voting procedure is a modified form of nominal group technique (NGT). NGT was developed as a means of problem identification and group judgement (Delbecq 1975) that avoids many of the problems involved in brainstorming or mind-mapping, where confident voices can dominate the agenda of the discussion (Rohrbaugh 1981). It essentially allows personal reflection on the importance and salience of topics without prejudice and bias emerging from groupthink. Secondly, it incorporates both opening up and closing down elements within the dialogue. New issues are raised and recorded, stimulating individual and group reflection processes, but the options for discussion are also narrowed through participant group reflection and voting to exclude those issues which are deemed by the group to be less important. This has significant advantages over attitudinal surveys or focus group methods which proscribe the framing of the issues and introduce researcher bias. This bottom-up issue framing method not only helps to reveal to researchers (and policy makers) which issues are of greatest importance, but also instils confidence in the process as being procedurally fair and transparent as the potential for external bias and coercion (in the Habermassian strategic dialogue sense) are reduced or removed.

Once this stage is complete and the clustered issues are transferred to the second board, the participants are given a second set of hexagons (yellow). This category is analytically referred to as actants (Williams-Jones and Graham 2003; Latour 1995) though it is termed 'people, objects, places and environments' for simplicity in a workshop context. This stage involves identifying the actors and affected groups, including future generations and the environment as well as technical and other artefacts. The identification and mapping of these elements is an important factor in the subsequent ethical deliberation process. The identification of disparate elements in an actor-network allows the participants an opportunity to evoke rich descriptions and reflections upon the role of technological artefacts in a broader social and moral context, by revealing linkages between human and non-human elements, between the natural and the artefactual.

When a series of actants have been identified through group discussion and recording it is necessary to then probe the ways in which these elements are related and the means through which they interact. The goal of the third stage is to identify the third and fourth categories shown in Table 6.2. Categories three and four are ostensibly representative of three dominant approaches to normative ethics – the consideration of action, behaviours and personal characteristics (representative of deontological and virtue ethics), and outcomes and impacts (consequentialism). By examining what the stakeholder actors do and the perceived consequences of their actions, participants can draw out a holistic picture of the decision-making context in terms of these contrasting ethical perspectives. It is important also to look at the ways in which technological artefact script agency: channelling actors to take specific courses of action or adopt certain behaviours, and conversely how actors can use technological artefacts in new and creative ways, thus rescripting the artefacts from their intended use by designers. An exploration of these relationships and negotiations between human stakeholders and the artefacts that link them is a crucial aspect of a rich and pragmatically grounded technology ethics.

Finally the last group, 'questions and issues' is aimed to assess and close down this actor network map, reflecting on which aspects of the discussion are worth carrying forwards, thus grounding the hexagon method in 'real-world' decision-making and reminding the participants of their progression throughout the discussion, providing opportunities for social learning within the deliberative process (see for example Bull et al. 2008; Schusler et al. 2003).

6.4.5 Linking the Different Elements

The advantage of the hexagon shape is the way in which different representations of relationships and linkages can be displayed through different configurations. Hexagons obviously tessellate across six sides rather that the four of a standard square or rectangular note which adds some flexibility to the visual style. Figure 6.1 shows how to represent the discussion diagrammatically, by linking the hexagons together. A ring of hexagons could represent a set of actants linked to a central ethical question, issue or behaviour. Here the idea is to show the actant/issue interaction, illustrating how different stakeholders are clustered around an issue, action or consequence with the divergent conceptual chains leading off from each stakeholder. A chain could represent a set of ideas that are linked either conceptually or chronologically and thus show a process of interactions. A cross-link illustrates how two different categories can be linked by a third, or the third joining hexagon can show a tangent where process chains diverge or coalesce. In each of these instances arrows can be drawn to illustrate the conceptual links between hexagons. These categories are not rigid. The aim is to encourage participant reflection on the issues and creative problem solving, so as long as the structure makes sense both to participants and to those third parties reviewing the outputs of the deliberation, then the method is considered a success.

Ring   Chain   Cross-link

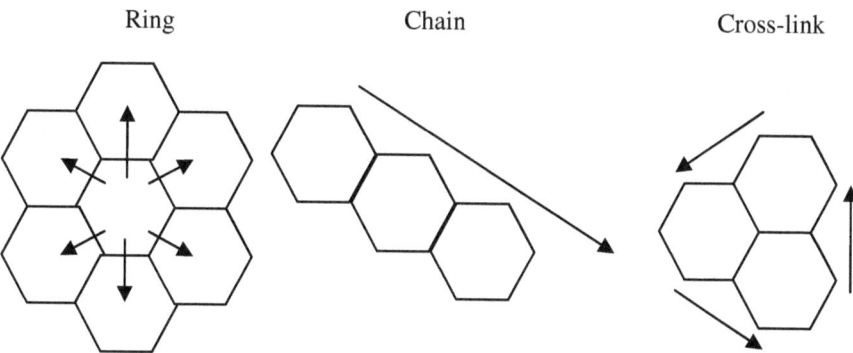

Fig. 6.1 Layout of the hexagons

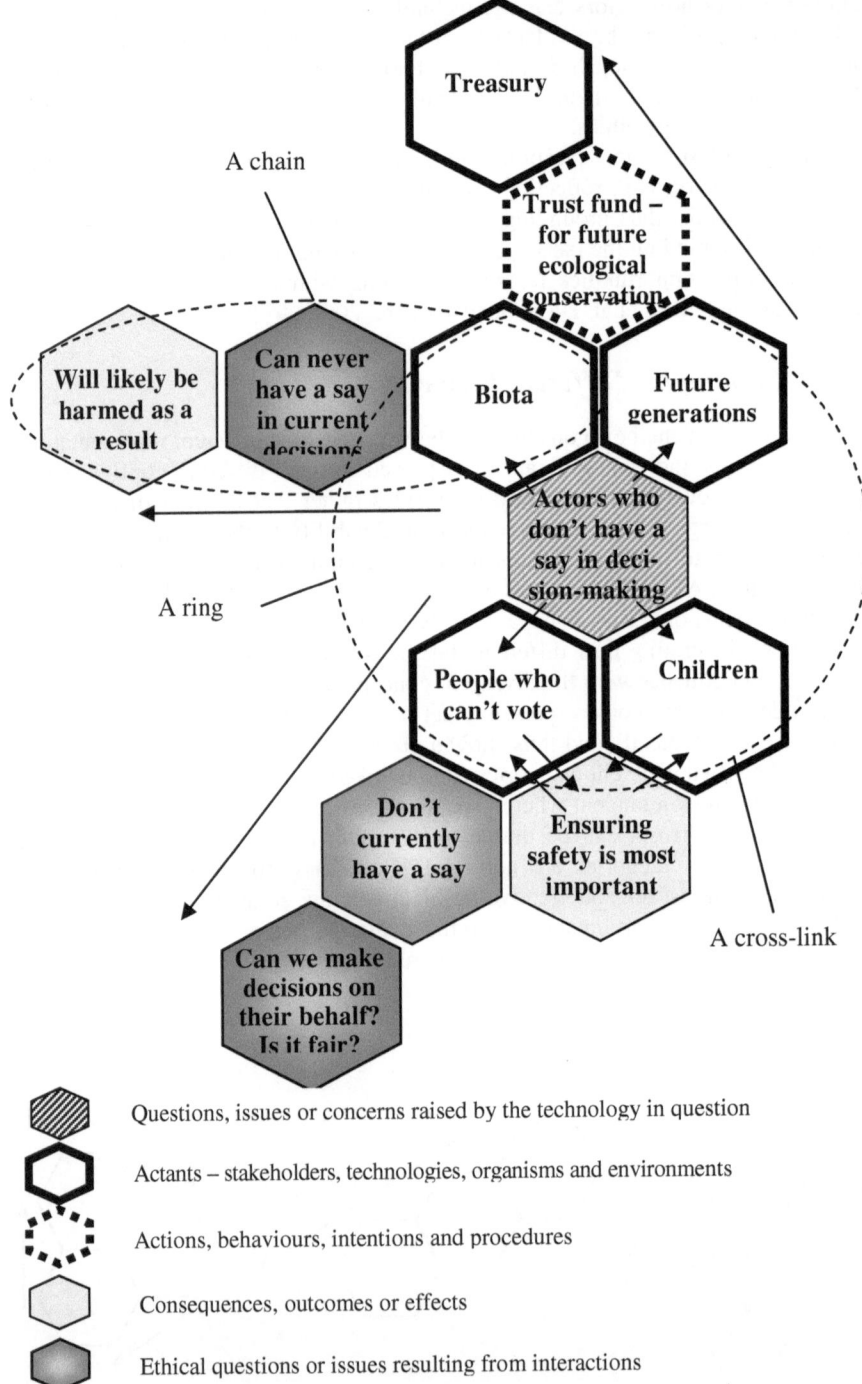

A chain

A ring

A cross-link

Treasury

Trust fund – for future ecological conservation

Will likely be harmed as a result

Can never have a say in current decisions

Biota

Future generations

Actors who don't have a say in deci-sion-making

People who can't vote

Children

Don't currently have a say

Ensuring safety is most important

Can we make decisions on their behalf? Is it fair?

Questions, issues or concerns raised by the technology in question

Actants – stakeholders, technologies, organisms and environments

Actions, behaviours, intentions and procedures

Consequences, outcomes or effects

Ethical questions or issues resulting from interactions

Fig. 6.2 Example hexagon map of radioactive waste management issues

Figure 6.2 shows a hypothetical example of the hexagon diagrams focussed on the issue of long-term radioactive waste management (real examples are discussed in the following section).

6.4.6   A Practical Summary of the Hexagon Method

As a practical summary of the hexagon method, it is estimated that the process should take around 2 hours in total:

- Discussion of information provided (10-15 minutes).
- Issue/question identification: each participant records approximately 3-5 of these, although it was made clear that more or less would be acceptable, each posted on flip-chart paper (10-15 minutes).
- Facilitator-led plenary discussion: highlighting further issues recorded by the facilitator on the blue hexagon notes. Thus two 'rounds' of issue and question elicitation, one that was 'from cold' and recorded by the participants themselves, and a second drawn out from the group discussions and recorded by the facilitator (10 minutes).
- Grouping or 'clustering' related responses: (either expressions of the same idea or linked by a common conceptual theme). Suggestions from the facilitator are put forward to the group as to possible groupings (if there were no forthcoming suggestions) and these are accepted, rejected or amended as a result of the ensuing discussion amongst participants (5-10 minutes).
- Voting: each participant is given a set of 5 sticky-backed dots, representing their individual 'stake' in the issues raised. They are then asked to cast their 'votes' on the salient issues (either singular issues where appropriate, or the grouped clusters) that they feel require further discussion. They are instructed to attach weight to the issues/questions accordingly, ranging from five dots on a single issue/question/cluster, to a single dot on up to five issues. (5-10 minutes).
- Actant identification: on yellow hexagons (5-10 minutes).
- Issue mapping: conceptually linking issues and actants (20 minutes).
- Identification of ethical issues related to the issue map (15 minutes).
- Second round of clustering: linking related ethical issues (10-15 minutes).
- Participant-led 'weighting' or 'voting' with 5 sticky dots on the ethical issues (5-10 minutes).
- Category labelling: selected for the following phase (5-10 minutes).

6.4.7 Real World Examples

In the following section, I present a series of outputs from the hexagon method, based upon a thematic analysis of the outputs across the three deliberative workshops. These are structured into two sections:

- Issues and actants identified
- Outputs of actant and issue hexagon maps

6.4.7.1   Issues and Actants Identified

The issues and question identification stage is designed to stimulate top-of-the-mind responses in a manner similar to opinion polls or attitude assessments. As such it is deeply influenced by the types of information presented at the start of the session, by local concerns that affect issue salience (such as in this case the decommissioning of local nuclear facilities and discussions over new build nuclear power in the case sites), and by external influences such as media coverage, the actions of NGOs, local protest organisations and political decisions, alongside myriad other external factors that shape the nature of public discourse. A thematic analysis of the outputs of the three workshops reveal a number of areas of enquiry that could be considered salient, though control for these external factors is not accounted for here. A number of issues were repeated and consensually agreed as important to the decision-making processes over radioactive waste management by participants, and are presented here to give some sense of the scope of the deliberative process. They are categorised into the following four groups:

Safety and Security
- Risk of terrorist attack on facilities, and concerns that terrorism is used as a smokescreen to promulgate a deep geological solution to radioactive wastes
- Fear over site security and potential theft of nuclear materials for profit or terrorist activities
- Safety of transportation of radioactive wastes from production sites to disposal facilities

Health and Wellbeing
- Prevention of accidents, especially Chernobyl style disasters
- Radioactive contamination of the environment and pollution of the biosphere
- Protection of future generations (both currently alive – children and grandchildren, and into the far future defined as 1000 years +)

Land Use and Technological Alternatives
- Advancements in future technology providing alternatives to disposal – such as partitioning and transmutation of long-lived radioactive wastes
- The transparency and communicability of technical and scientific criteria for site selection
- The provision of resources and incentives for scientific development of alternative energy sources, waste uses and waste management options

- Re-opening deliberation on the geological disposal option (as the CoRWM options assessment process had finished by this point)

Responsibility and Legitimacy

- Decision making authority – who makes the decision and on behalf of whom?
- Community involvement – what role do local communities play in decision-making both now and into the future, can they act as stewards for a waste facility
- Compensation – in what form, at what stage, who is offering it, and who will receive it?

What we is that these issues match those emerging from existing social research into the social and political dimensions of radioactive waste management processes (Marshall 2005; Rawles 2004; Cotton 2012; Mackerron and Berkhout 2009; Bickerstaff et al. 2008), in particular showing distinct similarities to perspectives articulated in psychometric risk analysis research on public responses to radioactive waste siting (Sjöberg 2003; Slovic et al. 2000). The issues prioritised by citizen-stakeholders encapsulated a desire to prioritise security measures to protect public safety in the face of terrorist threats and theft of radioactive materials; the prevention of nuclear accidents such as those seen at Chernobyl and Fukushima (though this later disaster had not occurred at the time of the workshops); and subsequently the consideration of alternative technologies for energy generation, thus linking energy production and waste management; the consideration of technical alternatives to deep geological disposal and issues of equity, fairness and decision-making – particularly the involvement of local community actors and the conditions under which they would accept such facilities in their local environment. In the wider context of PTA, what is important is the way in which these issues are arrived upon. To reiterate: through voting processes, these issues emerged as the most salient to the deliberative process from a bottom-up perspective of the involved stakeholders. The familiarity and similarity of the issues raised here and with broader population-representative social research into analytic-deliberative risk management in radioactive waste governance through psychometric measurement is suggestive of the socially robust nature of the approach to reveal the salient social and political dimensions of SECT.

Of equal importance is the identification of relevant actants. A second thematic analysis reveals listed groups of human and non-human elements. The results of the listing were integrated by joining them to the clustered issues that either affected these groups or upon which they had an effect. Over the three workshops the thematic list of 'stakeholders and affected parties' (as it was termed in the workshop) is compiled in Table 6.3.

Table 6.3 Identified stakeholder or 'actant' groups

Governmental organisations	Central Government
	Committee on Radioactive Waste Management
	European Union
	Health services
	Higher intra-governmental nuclear agency
	Local authorities
	Local government
	M.P. John Gummer (representing Suffolk Coastal district)
	Military
	NDA
	Nirex
	Police
	Schools
	Security Services
	The secret service
	Make up an independent body
Civil society stakeholders	Advocates of a community
	Children and civilians
	Scientists
	Human population worldwide
	The whole population
	Children
	Young people
	Host community
	The local population
	Future generations
	Terrorists (though these were recognised as politically illegitimate)
Objects and materials	Bricks and mortar (infrastructure)
	Chernobyl
	Electric power lines
	Municipal waste
	New energy efficient homes
	Nuclear power
	Nuclear Submarines
	Packaging
	Plutonium
	Sizewell power station
	Solar panels
	Spent Fuel
	Uranium
	Weapon materials
	Trident

Table 6.3 (*continued*)

Areas and environments	Australia
	France
	Habitats
	Land
	Other countries
	Scotland
	The local environment
	Animals and plants
	The UK
	England
	Suffolk
	Hartlepool
	Sizewell site
NGOs, independent bodies & businesses	Greenpeace
	British Nuclear Fuels
	British Nuclear Group
	Corporations and businesses
	Farmers and local food producers
	Foreign companies
	Greenpeace
	Pressure Groups
	Profit making organisations
	Shops and supermarkets
	The church
	The press
	Universities

The identified groups of actants presented in Table 6.3 illustrate the breadth of potential human and non-human elements involved in the landscape of radioactive waste management practice. The relatively broad scale of these groups effectively unbinds the ethical deliberation from consideration of predefined stakeholder groups, and by mapping together the relationships between these actants various synergistic relationships between them are more easily revealed than through a matrix structure. The types of actants identified is also important. With the exception of the group labelled 'terrorists', this method revealed breadth of the legitimate stakes in the policy process, which can help to frame both the 'who' of the engagement process for government or industry consultation on technology management, and also the 'what' – the types of artefacts and environments that should be considered. Though these category labels are fairly basic, and open to interpretation, they anchor the discussion to concrete elements of (a rather simplified) actor-network.

6.4.7.2 Outputs of Actant and Issue Hexagon Maps

The following discussion highlights some of the outputs of hexagon map discussions concerning two sets of issues mentioned:

- Safety, security and health
- Land use and technological alternatives

Brief sketches of discussion themes accompany the hexagon maps. These sketches or 'vignettes' are drawn primarily from the poster outputs, thematic evaluation of recorded audio and notes made during and immediately after the workshop discussions. Issues raised on the hexagons are presented in italics using the following notation:

- A – Actants
- B – Behaviours, actions, intentions and procedures
- C – Consequences, outcomes or effects
- E – Ethical questions or issues resulting from interactions
- Q – Questions, issues or concerns raised by the technology in question

6.4.7.3 Safety, Security and Health

These related groups of issues were deemed significant across all workshops. Often the issue of safety was framed in terms of twinned relationships between terrorism and the central governmental organisations involved in tackling them. One cluster emerged around the issue of terrorism (Q). Actants raised were central government (A), police (A), military (A), security services (A) and terrorists (A), which were all linked into a cluster of related groups (terrorists in the middle). Much of the discussion of UK's potential terrorist threats were framed in terms of a 9/11 style airborne attack on a nuclear reactor (hence causing a Chernobyl-style nuclear fallout scenario), or else the infiltration of a RWM facility and the theft of nuclear materials for radiological weapon-making purposes. This also linked with cross-cutting issues identified in common with environmental health and safety in areas such security of waste transport (Q) and safeguarding wastes for future generations (E).

Uncertainty over the timing and nature of an attack was a salient issue, identified as one that could affect the whole population (A), although no specific targets groups were suggested in any of the workshops. This issue was generally couched in the implicit ethical position that terrorism as a form of violence was morally reprehensible, and thus the actions of civil society actors in policy and security services to counter terrorist activity had tacit ethical justification. Participants in communities close to the Sizewell power station, questioned 'is the security good enough?' (Q), remarking that its vulnerability to terrorist attack had resulted in little change to actual nuclear operations – stating that the biggest impact that the War on Terror had had with regards to nuclear issues so far, was the closing down of the power station visitor's centre (B) (as a potential entry point for attack). This in turn sparked discussion of how information about nuclear issues had become less accessible; without that point of contact with the local community (and the UK population as a whole) locals may become more fearful (C) and distrustful (C) of nuclear power generally and RWM facilities more specifically. It was suggested that increased focus upon security issues nationwide limits the freedom and access to information on radioactive materials and thus distances local publics from scientific and technical bodies, which in turn leads to uncertainty, distrust and nuclear fear (C). Similarly though counter-terrorism was ethically justified, there was concern that terrorism is an excuse (C) to justify further expansion in the remit of their operations.

A second clustered group emerged concerning the relationship between RWM and other nuclear power-related risks. It is notable that no distinction was made between radioactive waste management, existing nuclear power, decommissioning of old reactors and new nuclear build. Across the workshops a range of inferences were made, linking issues about nuclear power (and in some cases nuclear weapons) to the issue of RWM; mirroring the findings of previous studies around risk perception and radioactive wastes that couch these technologies as dread risks (Flynn et al. 1990; Weart 1988). The Chernobyl (C) example was also used as an analogy for a RWM accident or contamination (C) event. Radiation was framed primarily in terms of the risk to children (A), young people (A) and future generations (A). There was evidence of what Douglas, Wildavsky and Dake (Douglas 1986; Wildavsky and Dake 1990) term *fatalistic* risk responses to the hypothetical situation of a nuclear explosion (C), with statements such as "if it [reactor or RWM facility] blew up we wouldn't know about it". A commonly expressed concern was that it would be those who survived and lived in the future that would bear the brunt of the costs, both in terms of economic clean-up and health risks from contamination, hence safety is most important (E) as an identified ethical issue.

Health concerns centred upon issues of leaking radiation from waste containers, decommissioned sites and power stations. Safety of nuclear technologies was linked to uncertainty (B), a concern over insufficient research into long-term radiation effects to the environment and human population and hence future generations (A). In some instances a broader theme emerged relating environmental impacts to healthy living and healthy lifestyles, relating power production from fossil fuels and nuclear against renewable energy such as wind, with the idea that the healthy body must exist within a healthy environment. Also, discussion centred on re-evaluating the concept of progress and development; challenging the accepted notion that nuclear expansion was necessary to meet continually rising energy demand. The primary ethical issue was responsibility (B): that current decision-makers and facility host communities would act as custodians of the wastes, guarding future generations from harm and ensuring long-term safety because they in particular don't have a say (C). Some participants suggested that our ancestors left problems (from technological advancement and resultant pollution) for 'us' and that we would do so in the future, thus it made little sense to try and safeguard them from the outcomes of inevitable technological progress. Others discussed how future technological developments could potentially neutralise radioactivity. Consequently, participants occasionally sought to re-open the issue of RWM option assessment, often expressing incredulity at the choice of the option of deep geological disposal. When and where this was accepted by the group, a general call for waste retrievability was expressed. A recurring theme was that community responsibilities for safeguarding wastes for future generations (B) contrasts with a sense that they would be better equipped to deal with them.

6.4.7.4 Land Use and Technological Alternatives

One area of relative conflict among participants surrounded doubt about deep geological disposal (Q), as some questioned 'what other options are available?' (Q). A minority of participants called for the reopening of the technology options assessment, while others trusted the legitimacy of the CoRWM-led option assessment process and were more accepting of deep geological disposal. This issue was repeatedly returned to throughout the workshops, alongside continual questioning of alternatives, such as disposal in outer-space (B) and immobilisation (partitioning and transmutation was mentioned). These ideas were popular because of their potential to reduce overall waste volumes or remove them from the natural environment altogether. However, with the outer-space option the issue of human error (B) and accidents (C) was raised (the space shuttle Challenger disaster of 1986 was mentioned in reference to this), and safety was considered paramount and hence argued to be sufficient justification to dismiss this option. The broader ethical implications of geological disposal were identified as an 'out of sight, out of mind' (E) problem, implying the sociocultural invisibility of risk (Beck 1996) whereby 'we can't see the risks' (C). This was construed as civil society's disregard for environmental safety, whereby the public presumes that once waste is secured below ground then it has been dealt with, and posited as a fundamentally dishonest (E) strategy. It was also identified as a global problem (E) and so comparisons were made with other environmental concerns at local levels such as municipal waste management and international levels such as climate change (C). The waste problem was often characterised as an international problem, raising questions about centralised international waste storage (B) and the exportation of waste (B) to other countries. An ethical debate about the exportation of waste ensued with two key points. Given that safety was held as the highest priority, some felt that an area with low population density but high levels of institutional control. Australia (A), the North York Moors (A) and Scotland - Mountains and Highlands (A) were all mentioned as specific areas that would be ideal for a RWM facility site, rather than the limited space (C) problem of a highly populated country, notably in England (A). This argument was countered by those that felt this was another example of the out of sight out of mind problem previously mentioned. Also, some recognised that NIMBY (Not-In-My-Back-Yard) (B) was a problem in any country and did not perceive that RWM could become safer, cheaper or more efficient in countries other than the UK. Centralised waste storage involving joint responsibility and compensation for the host country were discussed as potentially viable alternatives to a national strategy. Ownership of wastes (E) was an important factor in concluding against exportation, the burden of waste was broadly argued to be the responsibility of the producing country (or as some argued in the producing area). This intra-generational or regional equity problem (not their term) was recognised as an important aspect in accepting or rejecting a localised waste management facility siting proposal.

RWM was contextualised through comparison with pollution control across other industries. Some suggested adhering to a Polluter Pays Principle (although this exact phrase was not used) like other heavy industries and municipal waste management are obliged to do. Reduce, Reuse, Recycle, (B) emerged as a topic upon which to base the ethical justification for the elimination of nuclear new build. RWM was often explicitly linked to nuclear power as a continuous cycle of production and waste. Cessation of nuclear power would result in the reduction of waste and some argued that this should take priority. Waste's link to new nuclear power was recognised as being part of a problem of growing energy demand across the UK (and the world). The environmental benefits of lowering energy consumption were discussed particularly in terms of long-term radioactive waste reduction, as well as potential strategies for reducing demand such as replacing housing stock, investing in renewables, combined heat and power production (Micro CHP) and reducing waste heat from power stations.

RWM was framed by some in the broader context of a throw-away society (E) that was incapable of dealing with waste issues on a large scale. Climate change was a strong contextual factor, issues of energy efficiency and waste reduction, it was argued, should be addressed before proposals for new nuclear build. There was often evidence of a tacit assumption that waste management was intrinsically linked to new build and that the ethical principle of concern was that demand reduction should be the first priority.

6.5 Conclusions

The hexagon method presented here is something of a hybrid approach to ethical deliberation that draws upon existing stakeholder engagement tools to illustrate the interrelationships between heterogeneous elements of a socio-technical system, such as those involved in the management of radioactive wastes. The key issues raised by this method are accessibility and facilitation of effective decision-support. The simplicity of the method, listing individual actant categories and linking them together into conceptually contiguous groups belies the complexity with which the problem is evaluated by the participants. The intention is to visually display an Actor-Network, albeit a simple one, and in this task the method is broadly successful. The flat structure of an actor network is revealed in the linkages between the hexagons, and the method succeeds in providing a framework for relatively rich description of the relationships between heterogeneous elements. The advantage of the method also lies within its approach grounded in principle of generalised symmetry. When the different elements are broached, they are not hierarchically prioritised with certain actors at the top (such as stakeholders considered to be ethically motivated actors), and the technological artefacts and non-human biota considered to be inanimate and passive (and hence the object of the discussion).

Though this method proves useful in illustrating the socio-technical elements of the SECT in question, it does not present the means to evaluate the ethical issues inherent to its governance. Thus further tools are necessary to make implicit ethical issues explicit, and to weigh up their significance in light of a reflective process of moral evaluation; and the following chapter explores these aspects in greater detail.

References

Alexander, I.F.: A Taxonomy of Stakeholders: Human Roles in System Development. International Journal of Technology and Human Interaction 1(1), 23–59 (2005)

Anderson, J.R., Byrne, M.D., Douglass, S., Lebiere, C., Qin, Y.: An Integrated Theory of the Mind. Psychological Review 111(4), 1036–1050 (2004)

Andrén, M.: Nuclear Waste Management and Legitimacy: Nihilism and Responsibility. Routledge, Abingdon (2012)

Atherton, E., Poole, M.: The Problem of the UK's Radioactive Waste: What Have We Learnt? Interdisciplinary Science Reviews 26, 296–302 (2001)

Ausubel, D.P., Novak, J.D., Hanesian, H.: Educational Psychology: A Cognitive View, 2nd Aufl. Rinehart and Winston, New York (1978)

Avons, S.E., Phillips, W.A.: Representation of matrix patterns in long- and short term visual memory. Acta Psychologica 65, 227–246 (1987)

Beck, U.: Risk Society: Toward a New Modernity. Sage, London (1996)

BGS. Managing Waste Safely: Initial Geolgical Unsuitability Screening of West Cumbria. British Geological Survey, Nottingham (2010)

Bickerstaff, K., Lorenzoni, I., Pidgeon, N.F., Poortinga, W., Simmons, P.: Reframing Nuclear Power in the UK Energy Debate: Nuclear Power, Climate Change Mitigation and Radioactive Waste. Public Understanding of Science 17, 145–169 (2008)

Blowers, A.: Why dump on us? Power, pragmatism and the periphery in the siting of new nuclear reactors in the UK. Journal of Integrative Environmental Sciences 7(3), 157–173 (2010)

Blowers, A., Lowry, D., Solomon, B.D.: The International Politics of Nuclear Waste. MacMillan, London (1991)

Blowers, A.: Ethics and Decision Making for Radioactive Waste. Committee on Radioactive Waste Management, London (2006)

Blowers, A., Pepper, D.: The Politics of Nuclear Power and Radioactive Waste Disposal: From State Coercion to Procedural Justice? Political Geography Quarterly 7(3), 291–298 (1988)

Blowers, A., Sundqvist, G.: Radioactive waste management - technocratic dominance in an age of participation. Journal of Integrative Environmental Sciences 7(3), 149–155 (2010)

Bull, R., Petts, J., Evans, J.: Social learning from public engagement: dreaming the impossible? Journal of Environmental Planning and Management 51(5), 701–716 (2008)

Burgess, J., Chilvers, J., Clark, J., Day, R., Hunt, J., King, S., Simmons, P., Stirling, A.: Citizens' and Specialists' Deliberate Options for Mapping the UK's Legacy Intermediate and High Level Radioactive Waste: A Report of the Deliberative Mapping Trial. CoRWM PSE Working Group (June-July 2004)

Buzan, T.: The Mind Map Book: Radiant Thinking - Major Evolution in Human Thought. BBC Active, London (2003)

Carter, L.J.: Nuclear Imperatives and Public Trust: Dealing with Radioactive Waste. Resources for the Future, Danvers (1989)

Chilvers, J., Burgess, J., Murlis, J.: Securing Public Confidence in Radioactive Waste Management: Developing a Vision for a Process and Stakeholder Engagement. University College London, London (2003)

Coakes, E., Elliman, T.: Focus issue on legacy information systems and business process engineering: the role of stakeholders in managing change. Communications of the AIS 2(1), Article 2 (1999)

Collier, D.: CoRWM Final Evaluation Statement. Faulkland Associates, Oxford (2006)

CoRWM. Guiding Principles (2004) (accessed November 11, 2006)

CoRWM (2005), Why we need to consult, http://www.corwm.org/content-413 (accessed July 1, 2012)

CoRWM, Managing our Radioactive Waste Safely: CoRWM's recommendations to Government. Committee on Radioactive Waste Management, London (2006)

Cotton, M.: Ethical assessment in radioactive waste management: a proposed reflective equilibrium-based deliberative approach. Journal of Risk Research 12(5), 603–618 (2009)

Cotton, M.: Industry and stakeholder perspectives on the social and ethical aspects of radioactive waste management options. Journal of Transdisciplinary Environmental Studies 11(1), 8–26 (2012)

de Bono, E.: Six Thinking Hats. Palgrave, London (1985)

DEFRA, Managing Radioactive Waste Safely: Proposals for Developing a Policy for Managing Radioactive Waste in the UK: Department for Environment Food and Rural Affairs, The National Assembly for Wales, and the Scottish Executive (2001)

DEFRA, Managing Radioactive Waste Safely: A framework for implementing geological disposal. London: Department for Environment, Food and Rural Affairs, Department for Trade and Industry and the Welsh and Northern Irish devolved administrations (2007)

Delbecq, A.L., Van-de-Ven, A.H., Gustafson, D.H.: Group Techniques for Program Planning. Scott Foresman and Company, Glenview (1975)

Delbecq, A.L., VandeVen, A.H.: A group process model for problem identification and program planning. Journal of Applied Behavioural Science 7, 466–491 (1971)

Douglas, M.: Risk Acceptability According to the Social Sciences. Sage, London (1986)

Dunlap, R.E., Rosa, E.A., Baxter, R.K., Mitchell, R.C.: Local Attitudes Toward Siting a High-Level Nuclear Waste Repository at Hanford, Washington. In: Dunlap, R.E., Kraft, M.E., Rosa, E.A. (eds.) Public Reactions to Nuclear Waste: Citizens' Views of Repository Siting. Duke University Press, London (1993)

Flüeler, T.: Tools for local stakeholder in radioactive waste governance: Challenges and benefits of selected Participatory Technology Assessment techniques. Institute of Human-Environment Systems, Zurich (2005)

Flüeler, T.: Decision making for complex socio-technical systems: robustness from lessons learned in long term radioactive waste governance. Springer, Dordrecht (2006)

Flüeler, T., Scholz, R.W.: Socio-technical knowledge for robust decision making in radioactive waste management. Risk, Decision and Policy 9(2), 129–159 (2004)

Flynn, J.H., Slovic, P., Mertz, C.K., Toma, J.: Evaluations of Yucca Mountain: Survey findings about attitudes, opinions, and evaluations of nuclear waste disposal and Yucca Mountain, Nevada. Nevada Nuclear Waste Project Office, Nevada (1990)

Freudenberg, W.R.: Can we learn from failure? Examining US experiences with nuclear repository siting. Journal of Risk Research 7(2), 153–169 (2004)

Gemmell, C.: Long-term Radioactive Waste Management - Public & Stakeholder Engagement Consultation Document. Scottish Environment Protection Agency, CoRWM, London (2005)

Gregory, S., Satterfield, T.: Beyond Perception: The Experience of Risk and Stigma in Community Contexts. Risk Analysis 22(2), 347–358 (2002)

Grimstone, M.: Ethical and Environmental Principles: A Review of the Influence of Ethical and Environmental Considerations in the Formulation and Implementation of Radioactive Waste Management Policy. CoRWM, London (2004)

Gunderson, W.C.: Voluntarism and its limits: Canada's search for radioactive waste-siting candidates. Canadian Public Administration 42(2), 193–214 (1999)

Hare, M.P., Pahl-Wostl, C.: Stakeholder categorisation in participatory integrated assessment processes. Integrated Assessment 3, 50–62 (2002)

Hodgson, A.M.: Hexagons for systems thinking. In: Morecroft, J.D.W., Sterman, J.D. (eds.) Modelling for Learning Organisations. Productivity Press, Portland (1994)

Hodgson, A.M.: Hexagons for Systems Thinking. European Journal of Operational Research 59(1), 220–230 (1992)

IAEA, Ethical Considerations in Protecting the Environment from the Effects of Ionizing Radiation: A Report for Discussion (2002)

Kaner, S., Lind, L., Toldi, C., Fisk, S., Berger, D.: Facilitator's Guide to Participatory Decision-making. Community at Work, San Fransisco (2007)

Kemp, R.: The Politics of Radioactive Waste Disposal. Manchester University Press, Manchester (1992)

Kemp, R., O'Riordan, T., Purdue, M.: Environmental Politics in the 1980s: The Public Examination of Radioactive Waste Disposal. Policy and Politics 14, 9–25 (1986)

Kikuchi, T.: Temporal characteristics of visual memory. Journal of Experimental Psychology: Human Perception & Performance 13, 464–477 (1987)

Latour, B.: We have never been modern. Harvester Wheatsheaf, Hemel Hempstead (1993)

Latour, B.: La science en action: introduction à la sociologie des sciences (Science in action: introduction to the sociology of science). Gallimard, Paris (1995)

Law, J., Hassard, J.: Actor Network Theory and After. Blackwell, Oxford (1999)

Lidskog, A., Litmanen, T.: The Social Shaping of Radwaste Management: the Case of Sweden and Finland. Current Sociology 45(8), 59–79 (1997)

Litemanen, T.: Environmental Conflict as Social Construction: Nuclear Waste Conflict in Finland. Society and Natural Resources 9, 523–535 (1996)

Mackerron, G., Berkhout, F.: Learning to listen: institutional change and legitimation in UK radioactive waste policy. Journal of Risk Research 12(7-8), 989–1008 (2009)

Marshall, A.: The Social and Ethical Aspects of Nuclear Waste. Electronic Green Journal (21), 1–22 (2005)

McCutcheon, C.: Nuclear Reactions: The Politics of Opening a Radioactive Waste Disposal Site. University of New Mexico Press, Albuquerque (2002)

Miller, G.A.: The Magical Number Seven, Plus or Minus Two; Some Limits on Our Capacity for Processing Information. Psychological Review 63, 81–87 (1956)

NEA, OECD, The Environmental and Ethical Basis of Geological Disposal: A Collective Opinion of the Radioactive Waste Management Committee of the OECD Nuclear Energy Agency. OECD Nuclear Energy Agency, Paris (1995)

Nirex, Radioactive Wastes in the UK: A Summary of the, Inventory. United Kingdom Nirex Limited, Harwell (2002)

Novak, J.D., Cañas, A.J.: The Theory Underlying Concept Maps and How to Construct Them. Institute for Human and Machine Cognition, Florida (2006)

Novak, J.D., Gowin, D.B.: Learning How to Learn. Cambridge University Press, Cambridge (1996)

O'Hare, M., Bacow, L., Sanderson, D.: Facility siting and opposition. Van Nostrand Reinhold, New York (1983)

Peelle, E.: The MRS Task Force: Economic and Non-economic Incentives for Local Public Acceptance of a Proposed Nuclear Waste Packaging and Storage Facility. Waste Management 87 (1987)

Petersen, T.V.: Linked Arms: A Rural Community Resists Nuclear Waste. SUNY Press, Albany (2001)

Phillips, W.A., Christie, D.F.M.: Components of visual memory. Quarterly Journal of Experimental Psychology 29, 117–133 (1977)

Rawles, K.: Ethical issues in the siting of radioactive waste management facilities: the role of local communities. United Kingdom Nirex Limited, Harwell (2004)

Rohrbaugh, J.: Improving the quality of group judgment: social judgment analysis and the nominal group technique. Organizational Behavior and Human Performance 28(2), 272–288 (1981)

Schusler, T.M., Decker, D.J., Pfeffer, M.J.: Social Learning for Collaborative Natural Resource Management. Society and Natural Resources 16(4), 309–326 (2003)

Sjöberg, L.: Attitudes and Risk Perceptions of Stakeholders in a Nuclear Waste Siting Issue. Risk Analysis 23(4), 739–749 (2003)

Slovic, P., Flynn, J., Layman, M.: Perceived Risk, Trust and the Politics of Nuclear Waste. In: Slovic, P. (ed.) The Perception of Risk. Earthscan, London (2000)

Sundqvist, G.: Stakeholder Involvement in Radioactive Waste Management. Göteborg University, Göteborg (2005)

Weart, S.R.: Nuclear Fear: A History of Images. Harvard University Press, Cambridge (1988)

Wildavsky, A., Dake, K.: Theories of Risk Perception: Who Fears What and Why? Journal of American Academy of Arts and Sciences 119(4), 41–60 (1990)

Williams-Jones, B., Graham, J.E.: Actor-Network Theory: a tool to support ethical analysis of commercial genetic testing. New Genetics and Society 22(3), 271–297 (2003)

Chapter 7
Judging and Deciding

7.1 Introduction

The previous chapter describes the details and outputs of a hexagon mapping method to draw out a series of ethically informed questions and responses in relation
to a participant-constructed (albeit simplified) actor-network. This provides the
starting point for the consideration of individual reactions to the ethical content of
these actor-network relationships. The second phase of the workshop stimulates
reflection and discussion of individual judgements and intuitions that relate to
the ethical questions and ideas generated in the hexagon mapping phase of
the workshop. Critical to this process of reflecting on moral judgements is the stimulation of moral imagination in the participants, and this chapter details
the methods by which this can be achieved. Particular emphasis is placed upon the
role of imagery as a stimuli to moral reflection, and the operationalisation of
Rawls's reflective equilibrium concept as a means of evaluating moral judgements
in relation to a series of principles and vice versa.

7.2 Moral Imagination and the Elicitation of Judgements

The aim of the second phase of the workshop is to elicit a series of considered
moral judgements that arise in response to the ethical questions and concerns identified in the first stage. The making of moral judgements requires the elicitation of
a response in people, and hence the use of some form of stimuli to provoke a reaction and to draw upon the moral imagination of the participants. As discussed previously, there are difficulties with approaching ethics solely upon reason
(grounded in a rationalist perspective), when disconnected from intuition and
emotion. This is because when individuals are presented with a series of moral
rules, principles or theoretical frameworks to apply in making a decision, they can
nevertheless make moral mistakes. Such mistakes stem from what Werhane
(1999) calls moral amnesia – a habitual inability to remember or learn from one's
own and others' past mistakes and a failure to transfer that knowledge when fresh
challenges arise. Moral amnesia is caused by a lack of moral imagination.

M. Cotton, *Ethics and Technology Assessment: A Participatory Approach*,
Studies in Applied Philosophy, Epistemology and Rational Ethics 13,
DOI: 10.1007/978-3-642-45088-4_7, © Springer-Verlag Berlin Heidelberg 2014

Thus, I suggest that ethically informed decision-making must involve a careful balancing of real world context, evaluation, projection of moral standards and imagination. As Dewey, the pragmatist philosopher argues, ethics has a dramatic quality in the sense that it is concerned with *character* – the manifestation and interaction of personalities; with *plot* – creative descriptions and new narratives; and with *suspense* – the open-ended nature of moral debate (Dewey 1938; Caspary 2000). Dewey insists upon reflection in relation the intellectual habits through which we interrogate moral problems, because failing to do so will allow the metaphors that underpin our thinking and imagining to come to us mechanically, up to the point where we can no longer free ourselves from their influence upon us (Fesmire 2003), and hence a recurrent moral amnesia. I argue, therefore, that stimulation of moral imagination can alleviate this problem.

At the core of pragmatic ethical evaluation (particularly within a Deweyan vein), is a concern with the capacity of an individual with a highly developed moral imagination to perceive the nuances of a situation, challenge the framework or scheme in which an event, action or process is embedded and the capability to imagine how it might be different (Alexander 1993). In this way moral imagination can be defined as "a reasoning process thought to counter the organisational factors that corrupt ethical judgement" (Moberg and Seabright 2000). Moral imagination is posited as the key to developing sound ethical judgements in the reflective ethical mapping process because it facilitates (rather than replaces) moral reasoning. Moral judgements require cognitive reasoning processes and a measure of impartiality that are not merely imaginative. However, moral imagination helps one to disengage from a particular process, evaluate the situation and the mindsets which it incorporates, and think more creatively within the constraints of what is morally possible. Without this, one might remain mired in a particular situation, but without moral reasoning one could slip into fantasy (Werhane 1999, 2002). An imaginative ethics model would contrast with rational, empirical and calculative models of ethical decision-making that tend to involve the identification of alternatives, the estimation of advantages, disadvantages, costs and benefits; followed by the offsetting of these against each other in estimating which alternative is most advantageous or least harmful overall (McVea 2007). The advantage of reflective equilibrium as a model of ethical decision-making is in its ability to coherently balance these two aspects. The construction of moral judgements is stimulated by moral imaginative processes and the application of principles grounds the judgements in moral reasoning stemming from cognitive and analytical processes. The primary goal of this second stage of the workshop, therefore, is to find ways to stimulate ethical discussion of judgements and values in a way that is creative and stimulates moral imagination, followed by critical, theoretically informed reasoning applied to those judgements in an iterative hermeneutic circle.

In relating moral imagination to moral reasoning, one arguably important relationship is that between morals and aesthetics. Moral judgements are often claimed to be made by reference to general rules and principles whereas aesthetic judgements are made by reference to the particular features of what is judged. Therefore a moral matter involves acting towards some end whereas an aesthetic matter involves experiencing something for its own sake (Collinson 1985). However, within the field of moral psychology the growing popularity of social intuitionist models of ethical judgement such as the automaticity espoused by Haidt (2001, 2003), ethical judgement is akin in many respects to aesthetic judgement, in the sense that one reacts instinctively and emotionally to moral issues with a sense of approval or disapproval without having gone through an explicitly deliberative process of weighing facts and values. Judgements emerge complete within the moral consciousness with an affective valence. What Haidt argues is that moral judgements emerge instinctively, and attempts to then justify such positions involve a post hoc rationalisation of the judgement that is reached, rather than as result of going through sequential stages of philosophical reflection. They are in essence, to use the Deweyan term, moral habits. By stimulating an affective or emotional response to an issue, I posit that one can encourage participants to reach such judgements instinctively and then discuss them, formulating ways of explaining their position, though these explanations may be post hoc rationalisations of unconscious or perhaps more accurately, pre-conscious judgements. However, as a philosophical endeavour it is important not to stop there. The moral judgements espoused then present opportunities to record and critically evaluate the positions expressed. This has advantages for the empirical study of individuals' moral values, but more importantly they become the objects of an explicit deliberative process which reformulates such judgements in light of theoretically grounded principles. The seemingly reactionary, bottom-up elicitation of methods becomes carefully considered in light of common sense principles drawn from the wealth of ethical theory perspectives available.

7.2.1 Judgement Elicitation through Visual Stimuli

The question of how to stimulate and elicit moral judgements can be resolved by turning to methodologies within the social sciences. It must be noted that the intention is not merely the elicitation of values in the sense of drawing out innately-held attitudes, as if they were fixed, perfectly expressed internal representations of an individual's thoughts and feelings. The term elicitation is used here in a different sense, to imply a methodological tool designed to stimulate discussion and personal reflection, and hence encourage judgements to emerge through a discursive process. Various methods to stimulate such elicitation of affective responses have been discussed in the social science literatures, primarily facilitated by the use of visual, auditory or written stimuli. One method that has gained popularity in anthropology, sociology and cultural geography is an image-based research tool called the photo-elicitation interview. The method uses a photograph (or series of photographs), bringing them in to the process of a research interview, focus group or other qualitative data collection activity. By doing so, the photo is used as a device to frame participant responses, elicit affective, aesthetic or moral values, and to draw out rich descriptions from respondents in a way that talk or text alone may not. The subjective difference in responses between interviews using images and text and interviews using words alone lies in the ways we respond to these two forms of symbolic representation. As Harper (2002) suggests:

> "...this has a physical basis: the parts of the brain that process visual information are evolutionarily older than the parts that process verbal information. Thus, images evoke deeper elements of human consciousness that do words; exchanges based on words alone utilise less of the brain's capacity than do exchanges in which the brain is processing images as well as words... these may be some of the reasons the photo elicitation interview seems like not simply an interview process that elicits more information, but rather one that evokes a different kind of information."

Visual stimuli have proved effective in generating creative ideas, particularly when compared to verbal or text based stimuli (McFadzean 1997). This is because language can at times be a barrier to creative problem solving, and there is evidence to show that people when thinking creatively are more likely to use imagery than words (Proctor 1997). The use of picture-based stimuli can improve upon creative input to problem solving techniques when compared to text-based methods such as brainstorming or mind mapping (Vidal 2004; Higgins 1994); and the use of images, even when unrelated to the topic area, can stimulate useful associations and improve the creative aspects of problem solving or decision-making (Michalko 2006). The specific goal in using image-based methods is to develop a multi-staged process to elicit personally held beliefs and intuitive responses around areas relevant to the SECT in question, by stimulating deliberation and encouraging participants to express judgements about the ethical problems involved. Thus, the use of photographs and other images (cartoons, sketches, paintings) have been used as tools to expand upon questions or ideas in interviews and to allow participants to communicate dimensions of their lives, their environments and personal histories (Clark-Ibáñez 2004; Epstein et al. 2006); and can be of particular use in enhancing or complementing other qualitative research techniques (Hurworth 2003). Images have been used extensively either as an empirical data collection resource or else as symbolic representations and stimuli in qualitative anthropology and sociology (Prosser 1998). Their use is firmly established in participatory action research traditions aimed at empowering marginalised communities, especially those communities where text-based methods are unfamiliar or impractical due to language-barriers or differing levels of literacy (Heinonen and Cheung 2007; Smith and Emmison 2000).

Image-based research can complement text-based elicitation methods by stimulating imagination and visual memory, deepening the descriptions, values and associations discussed in the types of qualitative data collection that these workshops aim to promote. Satterfield (2001) in particular supports this ethos, critiquing the standard attitude assessment models prevalent in environmental valuation and technology assessment, arguing that speaking and thinking about different values, particularly ethical expressions of value, is ill-matched with the affectively neutral, direct question-answer formats standard to willingness-to-pay and survey methods. She asserts that morally resonant, image-based, and narrative-style elicitation allows new opportunities for respondents to express ethical values, articulating a broad range of non-cost and non-utilitarian values. These values are particularly pertinent to the group deliberations occurring within workshops structured around the reflective ethical mapping approach. In summary, I suggest that the use of imagery can be a useful means to stimulate discussion of ethical judgements because it provides a symbolic or proxy representation of an object, person or process that encourages reflection, discussion and a deeper consideration of the underlying issues than if text or discussion-based methods were used alone.

7.2.2 Developing an Image-Based Elicitation Tool

When using images as stimuli for ethical reflection, it is necessary to produce a broad palette of visual styles and a range of foci in order to stimulate personal and group deliberation and hence access the types of thinking that lead to personal moral judgements about the issues under consideration. In the workshops, single images were presented on a series of cards, each holding a simple descriptive caption. Examples of the captions and image themes are shown in Table 7.1. The choice of images is an important consideration, but range and breadth is the principal consideration. The images and captions are chosen to illustrate issues, objects or activities that are relevant to the case. They must be congruent with the types of information provided at the start of the workshop and the problem of the decision-making process. In short, they must be relevant to the case in hand, broad in their subject matter, and visually and discursively stimulating. These images are used as a device to identify the technical, social and ethical elements, thus expanding upon the deliberative exercises of the hexagon mapping phase. As the selected images are used as a framing device to structure the ethical discussion within the group, care must be taken to ensure that the bottom-up nature of the process remains intact. This is partly based upon the range of options and the capacity for individual choice. By allowing the participants opportunities to browse the images and to choose the ones that resonate with their personal reflections on the topic in hand, then the degree of individual autonomy and freedom from research-er/facilitator bias is reduced. When choosing images for display and selection by the participants, it is necessary therefore to sample such images from broad range of potentially stimulating aspects. One means to do this is to create a series of cat-egories ex ante, which encompasses the range of actants and perspectives identi-fied as related to the subject matter of the workshop. These can then be sampled (randomly or purposefully) to create image groups with an equal number image captions in each. In the workshops the range of images were categorised as fol-lows:

- Technological and design components – e.g. design schematics, maps, objects, formulae, engineers and scientists
- Environments and spaces – landscapes, urban, peri-urban and rural plac-es, local landmarks, architectural examples
- Symbols and designs – corporate logos, religious icons
- Famous individuals – e.g. politicians, religious leaders, celebrities
- People and relationships – young children, older people, relationships
- Emotive or unsettling – depictions of illness, wastelands,
- Non-human and biotic communities – rare animals, forests, oceans
- Conceptual and imaginative elements – future scenarios, future genera-tions, artistic representations of the other elements

It must be noted that this is not an exhaustive list, and other aspects can be cho-sen depending upon the situation, the policy context and the decision framing of the workshop. This aspect requires careful attention to the details and specificity of the case, so pilot testing of images is a useful means to select a broad array of stimuli. The value of the method is in the selection, discussion and application of these images by the participants themselves in relation to the topic under discus-sion, and so images must be evocative of a diverse array of themes. The bottom up nature of the process can therefore be further enhanced by participant led image selection and/or capture. For example local environments and spaces can be cap-tured by participant photographers or artists, thus enhancing the involvement of community stakeholders in the research/decision-support process.

In the workshop, images from these different categories can be displayed around the room in a gallery format. Participants are allowed time to view and re-flect upon the images prior to forming a group discussion. Before the discussion begins, the participants must each choose a selection of image cards that are placed in the centre of a board or flip chart, choosing the cards that represent is-sues that they believe important to the discussion and have particular relevance to the issues discussed in the previous phase. Participants examine the different im-ages and discuss the selection based upon the relevance of the images to the topics of discussion under consideration from the hexagon mapping phase. In the workshops this process was repeated for two different issues, thus allowing a breadth of images and discussion topics to be considered.

The use of images relates to the pragmatic goals of the workshop; to ground discussion and reflection of personal moral perspectives in something tangible that prompts or stimulates an affect-laden response. Images thus work as tools to aid memory and imagination, providing a point of reference upon which to move to more abstract philosophical concepts, and crucially providing methodological balance in the workshop by using a combination of text, image and verbal stimuli.

Some caveats remain. The process of image selection must be designed to ensure maximum group control and to foster equality amongst participants. They must agree upon a selection and post them up for further discussion. This has the advantage of encouraging group reflection on the purpose of the task through the transferral of individual image captions into a grouped selection; intended to counter the top-down aspect of pre-labelling the images. Group selection and organisation of the images adds a further level of subjective meaning, supporting the bottom-up problem framing necessary for the deliberative process.

7.2.3   Practical Summary of the Image Method

- Total time allowed: 1 hour 30 minutes.
- Group browsing of images, informal discussion and clarification of image themes (10-15 minutes)
- Group selection of images based on topic themes identified in previous hexagon mapping phase (10-15 minutes)
- First round discussion on emergent theme with most votes from the previous round. Facilitated small group discussion (6-8 participants), images placed down the left hand column of flipchart paper. Discussion is recorded by notation in the right hand column (20 minutes)
- Second round of image selection (including images already selected in the first round), repeat of step 3 for second theme (20 minutes)
- Final group plenary discussion of potential ethical issues emerging. Participants suggest what the ethical issues might be. These are recorded on a flip chart paper (20 minutes)

7.3   Practical Examples

In the following section I present a short sketch of some outputs from two of the workshops, giving examples of the different images that were chosen and the ways in which they were used to frame the discussions. Table 7.1 shows the discussion themes (drawn from the voting procedure of the previous hexagon method), and the caption labels of the images chosen.

Table 7.1 Chosen images representing safety and security

Workshop 1 – Leiston	Workshop 2 - Hartlepool
Trust and safety - discussion 1	**Fear and danger - discussion 1**
Coastline	Deep geological disposal
Deep geological repository	Deep geological repository
Dirty bombs	Farmland
Future generations	Future generations
High level waste	Future society
Intermediate level waste	Hartlepool town square
Radiation poisoning	Heavy industrial areas
Rail transportation of wastes	Nuclear fuel reprocessing
Road transportation of wastes	Nuclear site security
Scientists and technical experts	Nuclear weapons testing
Sea level rise	Radiation poisoning
Sea transportation of wastes	Road transportation of wastes
Suffolk coastal region	Terrorism
The prime minister	The prime minister
The world	Warfare
Compensation - discussion 2	**Local issues & public opinion discussion 2**
Compensation/community benefits package	Climate change
Conservation	England
Journalists and the media	Ghost ships & Local councils (linked)
Lakes	Hartlepool local M.P. (Ian Wright)
Marshland	
Plants and trees	Journalists and the media
Sites of historic interest	Local businesses
Sites of special scientific interest	Nuclear protest
Teachers, schools and education	Onshore wind power
Teenagers and young people	Rioting
Woodland	Teenagers and young people
	The public
	The World
	Voting

7.4 Some Emergent Themes

Brief sketches of the discussions are outlined below where there were overlapping issues emerging in both Leiston and Hartlepool workshops. Emergent themes are discussed with reference to the relevant images listed in Table 7.1.

7.4.1 Safety, Hazards and Risk

The most notable aspect of the safety issue was that it was primarily framed in anthropocentric terms, i.e. towards protecting communities living close to radioactive waste facilities, rather than upon environmental or ecological protection. Inference to safety issues was drawn from a series of human failures either technical and engineering errors, or operating errors, and participants in the Leiston workshop made reference to the images on rail and road transportation of wastes, whereas in the Hartlepool workshop they made reference to heavy industrial areas. The risks of technical and system error were linked with a lack of information provision to local communities with existing radioactive wastes. This was prompted, in part, by highlighting Chernobyl as a lesson in human error-related nuclear catastrophe, and hence radiation poisoning in both the Leiston and Hartlepool workshops. In Hartlepool there was also a suggestion that scientists sought to control a technology that is inherently dangerous and unpredictable. Additional risk factors were identified, such as waste transportation at sea (with analogies to oil tanker disasters, the Hartlepool 'ghost ships' and the MSC Napoli off the Devon coast), and transportation was discussed as one of the key risk factors in finding a suitable site. The waste management issue was also related back to the broader 'safety culture' in the UK; specifically to how risks are managed by technical experts and how the public has a lack of trust towards these authorities, with reference to nuclear site security. Also the issue of human risks was generally considered to extend beyond human error to the possibility of sabotage and terrorism in Hartlepool and dirty bombs in Leiston, prompting concern over the safety of radioactive waste management facilities. The nature of terrorist activities was also seen to be changing, with terrorists no longer concerned for their own personal safety; arguably making them more dangerous if personal risk was not a factor in their actions.

A number of other common themes were raised, specifically regarding the uncertainty involved in managing the wastes over long time-scales, and so future generations were chosen in both workshops and the importance of 'getting the science right' was stressed – incorporating knowledge about (for example) climate change and coastal erosion in evaluating waste management strategy safety. As such, 'external' risks do not fall into the category of 'human error' based safety concerns. At times participants expressed distrust in scientific and technical authority and at others, asserted that adequate scientific evaluation was a prerequisite to guarantee long-term public safety.

7.4.2 Compensation and Community Decision-Making

This discussion around the ethical issues of compensation/community benefits package was, to some extent, framed in terms of the relationships between corporate interests and communities. To some participants, the waste issue stemmed from industry and thus the liability should be owned by the producer, as the majority of wastes are produced by profit generating nuclear power stations. Thus, the idea of a compensation/community benefits package was framed in terms of individual and community rights being infringed by corporate actions involving pollution. To some, the issue involved an explicitly ethical standpoint, an issue as fundamental as environmental and community protection should not be decided on the basis of further material consumption, i.e. buying or building a new set of material goods does not outweigh the risks and costs (both economic and environmental) of waste management. In the Leiston workshop a range of natural environment images were selected and referred to: Lakes, Marshland, Plants and Trees, and Conservation. To others, the question of the ethical validity of a compensation/community benefits package came down to the manner of administration, in particular the stage in the process at which it was offered to the community. If it is offered before a siting proposal is made then this was deemed to be bribery, and only when administered after site selection could it be considered compensatory. The established themes of waste reduction resurfaced in the discussions; avoiding material consumption and contextualising the waste issue in broader terms of reducing consumption locally and globally, with a general rejection of the idea that economic measures could ever morally compensate for environmental degradation.

The issue of community roles in decision-making was raised in the Hartlepool workshop. Little faith was expressed in the power of local people to influence decision-making processes and there was broadly a consensual distrust in the authority of local councils and their competency in decision-making, and also in national level consultation processes. Participants felt that despite consultation, final decision-making power would rest in a top-down ministerial decision, with reference to The Prime Minister, and this undermines any partnership-type role for local people. Parallels were drawn with recent government consultations on the future of the local hospital, which all participants felt had been a waste of time, with local viewpoints being ignored in decision-making. This lack of faith in consultation and community partnership was also seen to undermine an adequate ethical assessment of the issues, as ethics was seen to be absent from centralised decision-making processes. The status of 'the local people' as a homogenous group was also in dispute. Participants recognised that there was no consensus among them about who should represent a community, given their lack of trust in local councils or how this representative could stand on behalf of their interests. This related to the issue of compensation and community benefits, as without consensus on what this should look like, it would make an inadequate measure to alleviate the risks of RWM in the local area. Without adequate community representation it was noted that protest actions and even rioting would become a problem, though even this was considered morally preferable to a technocratic, top-down decision from central government that the local community would be unlikely to support.

7.5 Reflections on the Method

These brief sketches of discussion themes, give an idea as to the use of image captions in providing a contextual frame for deliberation to develop. They help the process by maintaining topic focus throughout, and hence encourage the participants to 'stay on course' in reaching the decision-support portion of the workshop in later phases. In practice, participants tended to utilise the images in different ways, in some cases to explain or justify a particular point they wished to make by referencing the image caption whilst explaining their argument, pointing to or gesturing at the images when speaking about particular issues, or else they discussed the choice of image that one another had selected, thus strengthening the dialogic quality of the process. There is also evidence that these images have an effect on stimulating moral imagination, where images are used as anchoring devices - reference points upon which to justify specific responses to issues raised, and encouraging them to consider a range of different viewpoints and perspectives. For example:

> Leiston participant: I put [former Prime Minister Gordon] Brown's image up there because people like him, the likes of him, they can change their mind just like…. I've got to quote this, the people in England would like a referendum and he says "no", so there's your power struggle there, he's the one who'll decide, it doesn't matter what you say. That's my problem with the top.

Or to give another example:

> Hartlepool participant 1: I chose the image, it wasn't about Christchurch [a local church in the town centre] it was just an image of Hartlepool and …

> Hartlepool participant 2 : ….local issues?

> Hartlepool participant 1: local issues, yeah, I'm a great believer in the number one priority, all I've said today is look after your own look after the people on your door step. And the other image is about the future, what is the future going to be? It's a very uncertain place. And the decisions we've been asked to make is really our problem and it's difficult for us. If you go to various meetings and the nuclear industry will tell you it's a community's waste. It's not a community's waste its

British Energy's waste or it's industry waste, therefore, just the uncertainty there for the future and the uncertainty about making a decision on where we go and who's responsibility it is

Hartlepool participant 3: Is that what they say, "it's a community's waste"? If it's a community's waste why don't they pay for us to accept it?

Though by no means a comprehensive qualitative analysis, these brief exchanges reveal some of the potential benefits of using images in structuring dialogue, advancing the discussions of the previous hexagon mapping phase by providing concrete visualisations of the ethical issues under consideration. As shown in these utterances, the imaginative stimulation of these visual representations provides a particular kind of discursive space through which participants can question motives, examine trust relationships and make judgements about individuals, organisations and the actions that they take. This helps to move discussion towards the consideration of specific judgements in relation to these issues in the following workshop phase.

7.6 Eliciting Judgements Using a Charrette

Image-based framing of the discussions aims to identify areas in which imagination could play a part in encouraging individuals to make judgements about the issues under consideration. It then becomes necessary for them to explicitly state what these judgements are, and to make this transparent to the group and to third party evaluation of the process. The recording of judgements can then be elicited through the use of listing methods, whereby judgements can be sequentially recorded and discussed by participants. One such method of listing is termed a charrette (origin from the French for 'cart' or 'chariot' – in reference to student architects at French design schools working up to a deadline, whereby a cart or charrette would be wheeled amongst them to pick up the work for review. Those still working to apply the finishing touches were said to be working *en charrette*, in the cart). Charrettes are structured deliberative methods conducive to collaborative development of scenarios, and used in planning, design and group problem solving activities. They provide an iterative review process of idea development and refinement, involving rounds of discussion in small groups with addition of new ideas in each round. The key facets are the emphasis on group working, iterative development of ideas and the imposition of a time limit on discussion and design activities.

In the previous image method, problems are framed in terms of emergent ethical issues. Once this stage is complete, the charrette aims to allow groups to discuss one issue for a fixed period of time and through discussion draw out individual judgements about the ethical issues presented – 15 minutes for the first round, then 10 minutes for each subsequent round. After each allotted time period the groups swap and discuss the second issue, the third and so on, until all issues have been discussed by all groups. At each stage the participants record the judgements on sticky notes (using a specific colour - in this case yellow) and put them up on a board, posted sequentially as the discussion progresses. A facilitator can help to record these and thus keep the flow of the discussion going. At each successive round the new group can only add new judgements, they cannot change or amend anything that had been discussed before. At each swap, one member of the previous group outlines the main points of their discussion with the new group before moving on to the next issue.

Throughout the process participants are instructed to frame their expressed judgements in terms of a specific normative or value statements such as "I believe we should do this", or "an institution/actor ought to do this", "this action is right," or "this policy is unjust". The ethical judgements in question are intended as subjective statements with a normative value, which can later be assessed in relation to a series of ethical principles in order to stimulate a reflective equilibrium. The use of these statement forms forces participants to consider basic moral binaries and to put forward judgements as statements of intent. It was made clear that the point of the exercise was not to criticise or comment upon individuals' personal beliefs, but to consider how they fit into a wider pattern of moral principles and see where the relationships lie. Following the completion of the charrette, the post-it notes are reorganised by clustering them into contiguously related categories and weighted according to participant views on their importance for evaluation, using the nominal group technique seen in previous rounds.

7.6.1 Practical Summary of the Charrette Technique

- Total time, approximately 1 hour, 15 minutes.
- Divide participants into groups, give each individual a set of post it notes.
- Set ethical 'topics' emergent from voting process in previous stage.
- Each group discusses the first issue recording judgements sequentially (15 minutes).
- Groups switch topics – one participant describes outcomes of first round (5 minutes).
- Second round of discussion and judgement recording (10 minutes).
- Groups switch topics again – one participant describes outcomes of second round (5 minutes).
- Final round of discussion and judgement recording (10 minutes).
- Plenary discussion of outcomes (20 minutes).

7.7 Examples of Ethical Judgements

Below I give some examples of the groups of judgements that were contiguously related around common themes of ethical issues in relation to the long-term management of radioactive wastes, using the issue of compensation/community benefits to highlight the types of judgements and intuitions that emerged.

The issue of compensation is crucial to the management of radioactive wastes. Current UK policy strategy for long-term radioactive waste management involves a process of providing community benefits packages (universally described as either compensation, or bribery by participants in the workshops). The notion of when compensation or benefits should be provided, as well as the form it should take and who the beneficiaries should be, are key ethical issues that were explored in the workshops. In the Leiston workshop in particular, the issue of compensation was central to their understanding of radioactive waste management facility siting as an ethical issue. Compensation/community benefits as a title category emerging from previous rounds of discussion was then subdivided into linked subcategories of judgements related to personal gain and greed, reducing energy consumption, costs, and siting. Examples of the written judgements emerging from the charrette procedure are displayed below:

7.7.1 *Community Benefits*

- Community benefits should be ongoing
- Benefits should be distributed globally, not just locally
- Nuclear gives clean air – less CO_2. We all benefit and individuals should accept this
- Compensation/benefit is a must
- Compensation should benefit both the individual and the community
- Compensations should include insurance assistance in the case of accidents
- It must be a community benefit – ensuring that all affected have access to rewards
- Personal gain/greed
- Bribery is just another form of control and corruption and must be avoided
- Unfortunately, people will usually think of their own personal gain over the greater good of all
- Bribing communities to take on nuclear waste does not sort the long-term complex problem of waste, it only satisfies the short-term greed of a few individuals
- Human greed will be the downfall of the entire planet, we need to stop taking
- Compensation is just a way for large organisations to make 'the small people' change their views and opinions. It is BULLYING PEOPLE!

7.7.2 *Reducing Energy Consumption*

- Compensation should be aimed at reducing overall energy consumption for the good of the planet

- I think if compensation is to be used, it should be given to businesses and individuals who reduce their consumption
- Costs
- How can compensation in whatever form even compensate the community who suffers nuclear disaster
- We shouldn't pay twice – as a consumer of nuclear electricity and later as a tax payer funding waste management

7.7.3 Siting

- Finding a site for a waste dump should be done solely on geological grounds
- It should not be in my back yard!

Across these examples there are clear themes emerging. Firstly, the institutions that would be providing the compensation remain nameless, it was unclear to participants who would be compensating whom, and so they remained distrustful of organisations that might provide such incentives. There was also clear disparity amongst participants about the ethical values and motivations held by those that offered compensation/community benefits packages. The two primary themes were, firstly, a position that compensation was unconditional bribery reflecting immoral societal 'vices' such as corporatism, greed and excessive materialism; and secondly, a somewhat more pragmatic approach that community benefits were a just exchange for the acceptance of new environmental risks. For some, the issue of siting a waste facility was only considered ethically valid when based primarily upon objective scientific criteria (i.e. the geological suitability of a location) without any form of incentive. For others, they unconditionally would not accept waste in 'their back yard', implying a NIMBY or more accurately, NIABY (not-in-anyone's back yard) position whereby responsibility for waste management should never be held by individuals to bear excessive technological risks for a power generation source that they did not personally support. In contrast, some participants expressed what could be considered utilitarian positions, asserting that national safety is the primary concern that overrides all other community-based concerns. Some even advocated stronger centralised institutional control to 'force' planning for RWM facilities into geologically suitable sites if the techno-scientific 'safety case' (their term) was strong enough to justify this. Some saw the benefits of RWM facility builds and new build nuclear power locally, in terms of local employment and wider benefits from CO_2 reduction and hence climate change mitigation, but had no specific requests in terms of local benefits. Issues of cost were raised with a concern about having to essentially pay for the waste twice, firstly as an electricity consumer and secondly as a tax payer, and the moral implication being that this would hurt the poorest the most.

From these sketched examples, the elicitation process is shown to produce a fairly diverse range of judgements and intuitions around specific categories of ethical issues. To broadly categorise the tenor of these responses, the judgements tended to fall into the following three groups:

1. Express dissatisfaction with current institutions, behaviours, policies and practices
2. Suggest potential strategies, policies or practical recommendations that should be carried out
3. Express concerns, personal values or comments on broader public and moral social values.

These judgements and intuitions are variably prescriptive and descriptive depending upon the context in which they are put forward. At times normative ethical judgements are stated, implying specific actions should be taken. At other times, descriptive and reflective judgements about human behaviour, policies and actions are expressed. It is important, therefore, in subsequent phases of the workshop to examine the diverse array of judgements and their interaction with similarly diverse principles, in order to assess their interaction might influence the quality and ethical substance of the deliberation, and how judgements are contextualised as courses of actions that branch out as different principles are applied in action.

7.8 Applying Principles

In practical terms, once the judgements have been elicited and recorded and a break has ensured, a second stage of the reflective equilibrium model is initiated. Participants are presented with a list of pre-selected principles which is then placed to one side where all can read the definitions, and a further selection of square sticky notes (green) is stuck to the side of a display board each containing a single word category label to represent each principle. The ethical principles used in the workshops were identified as a list drawn from an examination of the literature on principlism in applied ethics. Examples include the aforementioned Beauchamp and Childress (2001) principles – Autonomy, Utility, Beneficence and Non-maleficence, which were used alongside others idenitifed from academic sources (Kaler 1999; Schmidt-Felzmann 2003; Grassian 1981; Rachels 1993). The initial list included the following:

- Autonomy – The right of individuals to make free and informed choices
- Utility – The greatest good to the greatest number
- Fairness – Treating everyone equally. Addressing the imbalance between those with more and those with less
- Honesty – Being truthful, not telling lies or misleading others
- Fidelity – Keeping agreements and upholding promises, contracts and oaths
- Beneficence – Helping others and doing good

- Non-maleficence – Not harming other, avoiding wrong-doing
- Duty – the golden rule, 'do unto others as you would have them do unto yourself'
- Justice – Individuals receive that to which they are entitled. Good actions are rewarded and bad actions punished.

This list is not intended to be exhaustive. Alternatives can and should be identified where appropriate to the case; and the reflective ethical mapping process encourages not only the expansion of this list of principles, but also a re-evaluation of the meaning and context of these principles in relation to context-specific reflection on the case in hand. Though guided by the philosophical grounding of a principlist approach, the opportunities for amendment provide a degree of bottom-up context validity. It must be stated that from the workshop process, a number of new principles emerged from the participants' discussions along with accompanying definitions:

- Transparency – the need to be not only honest but forthcoming about decision-making processes
- Sustainability – The long-term balance maintaining the future environment. The need to survive.
- Precautionary Principle – trying to reduce potential harm to people and the environment from dangerous technologies.
- Legal justice – the laws of the land, enforceable in court
- Natural justice – laws of nature, higher than government legislation
- Inherent value – that all beings are valuable in and of themselves

Each principle must be presented with a concise definition (such as the ones shown in the above list). However at any point during the discussions, participants are encouraged to question these definitions and make amendments based on whether they seem relevant to the case. They are also encouraged to suggest other principles that have bearing on the problems that they identify. Like previously mentioned methods such as the ethical matrix, this approach is primarily principlist, based upon Beauchamp and Childress (2001) dialectical approach; achieved by placing the clusters of identified judgements from the previous stage and instructing participants to discuss these judgements in relation to principles. The recorded judgements from the previous phase on yellow sticky notes are arranged on the board, and they are then asked to consider the range of principles that are described on the sheet on the wall, and to decide between them which principles the judgements were invoking. This requires careful facilitation – encouraging participants to choose the ones that they think are relevant and to explain where the link lies and why. Care must be taken not to criticise choices of principle selection, and even those that may not intuitively link together can nevertheless produce surprises of moral reasoning amongst participants.

By then placing the relevant principle (for example on green sticky notes) next to the judgement they are asked to join principle and judgement together and discuss the implications of applying the principle to the judgement, i.e. what would be the logical outcome of applying the principle in a course of action, or what strategy would be implied by following the principle. These are then annotated with additional contextual factors that emerge in the discussion of the outcome of principle-and-judgement comparison. These broader contextual factors are recorded on a third coloured sticky note (in the workshops pink was used). By arranging these together, using a multiple branching system of judgements (yellow), principles (green) and other contextually relevant factors (pink) they work to produce a conceptual map that is representative of coherentist ethical reflection in the wide reflective equilibrium approach. The judgements can branch out into new territory when new principles are applied, and similarly the principles themselves can be compared by drawing relevant practical examples and the discussion of moral questions and contexts emerging through group deliberation.

7.8.1   *Practical Summary of Reflective Equilibrium Technique*

- Arrange elicited judgements into related groups with participant input (5-10 minutes)
- Introduce range of ethical principles, discuss alternative definitions and new principles not currently included (10-15 minutes)
- Discussion and principle selection, application and reflection (10 minute cycles – overall 45-60 minutes)
- Plenary feedback and discussion of reflective equilibrium map (10-15 minutes)

7.9   Example of Reflective Equilibrium Technique in Practice

In the explanation below, and in figure 7.1, I present a reflective equilibrium-based conceptual map that draws on the issue of risk in relation to nuclear power and radioactive waste management. At the centre of figure 7.1 is the category label 'nuclear risk '. Within this cluster were a range of judgements that related to concerns over cancer risks (for example drawing on the Chernobyl catastrophe, and cancer clusters near nuclear power stations), concerns over the destructive ethos of nuclear technologies, and the concept that not all of the risks are 'real' in the sense that some are pursued as legitimate and others are not. Using the sticky notes, participants branched out the ethical issues into three main trajectories. The first concerned 'keeping the power on' similar to the expression 'keeping the lights on', terminology used as a succinct descriptor of a (perhaps moral) imperative to bridge a growing energy gap between the decommissioning of nuclear facilities, reducing overall supply and projections of constantly rising energy demand (Patterson 2007; Makansi 2007). Secondly, there was a group of issues related to cancer, particularly cancer clusters and the pervasive, invisible risks of radionuclides. Thirdly, the branched judgements concerned the issue of honesty about risk, who is responsible for researching nuclear risks and communicating risk information to the public. I've broken down these issue groups into three sections below:

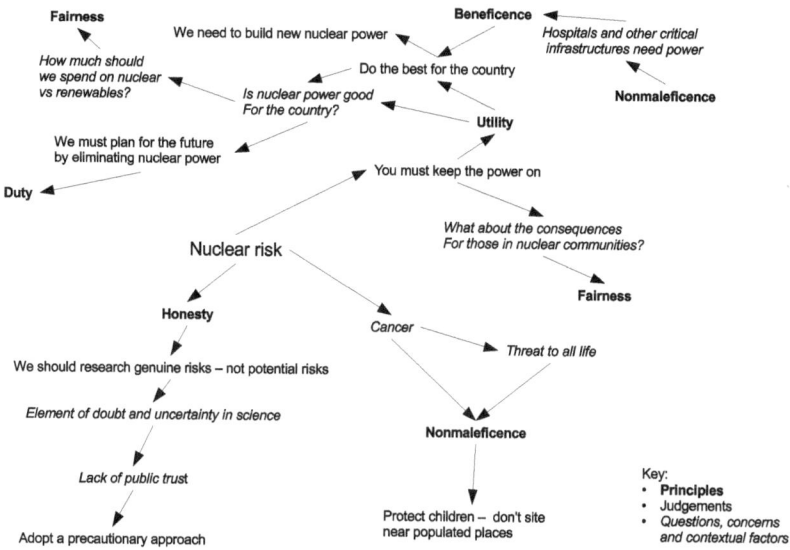

Fig. 7.1

7.9.1 Keeping the Power On

The principal ethical motivation for new nuclear build and hence continued radioactive waste production, was construed as a position on keeping the power on, in reference to concerns over blackouts affecting vulnerable infrastructures such as hospitals. The framing of these judgements was construed as one of utility, itself branched into two lines of ethical thought. In one exchange, the participants' discussion evoked concepts of welfare utilitarianism (Goodin 1995), in that (one of the few) positive goals of nuclear power is to provide the necessary conditions in which to live comfortably, i.e. by reducing the vulnerability of the ill and injured in hospital from further suffering under conditions of energy scarcity/blackouts. This welfare issue was discussed in relation to the twin principles of nonmaleficence and beneficence, thus the ethical principle of utility in relation to nuclear technologies was contextualised in relation to notions of harm reduction and welfare promotion. In another linked exchange, the concept of utility was applied to the notion of the 'public good' – whether nuclear powered electricity was beneficial or harmful to the UK population as a whole. Through the counter principles of duty and fairness (invoking a tension between deontological and egalitarian ethics on one side and utilitarian ethics on the other), the group decided that the elimination of nuclear new build was desirable, because nuclear power expenditure was construed as having a zero sum relationship with renewable energy technology investment, and thus was unfair. They posited that the UK had a duty to phase out nuclear power and instead invest in renewables. Though a preference for nuclear opposition was clearly expressed in these exchanges, participants realised that to take a purely utilitarian position would lead to a decision to support new nuclear build. We therefore see some evidence of social learning through the process of deliberation, by comparing principle perspectives and how they logically entail different courses of action. This encouraged participants to engage in ethical reflection, clarifying the terms by which they made policy choices in relation to the technology.

7.9.2 Cancer Risks

The second branch concerned the issue of cancer. This was discussed in relation to concerns over cancer clusters in areas close to nuclear reactors (and linked to the third branch around the question of honesty). In these exchanges the principle of non-maleficence was discussed, the notion that 'first do no harm' should be the guiding principle around waste management and the development of nuclear energy policy. It was recognised that there were different scales to the harms that could occur. Chernobyl was mentioned as an example of the global scale of harms resulting from nuclear risks, and the localised risk of cancer clusters around nuclear sites was posited as a local harm. At the heart of the ethical principle guiding nuclear technology development and implementation was the protection of children (construed as future generations in this context). The siting of waste facilities near populated areas, particularly in areas close to schools, was considered morally undesirable, and should be a primary criterion of nuclear waste repository siting.

7.9.3 Risk Communication

The third branch concerned honesty in relation to the concept of risk communication. Participants questioned what counted as real risk and what didn't, in relation to the preceding discussion on cancer clusters around nuclear waste sites. It was agreed that there was not enough information coming around the 'real' risks of nuclear power, not simply because of information availability, but also a lack of independence. Risk information from nuclear industry sources was not trusted, independent scientific information was not accessible (concern over the suppression of independent scientific research into nuclear risks was also mentioned). Because of the lack of public trust and scrutiny of the nuclear industry, the concept of the Precautionary Principle was mentioned as a new addition to the list of principles displayed. Though there was some facilitator led discussion as to whether this was an ethical principle per se, participants agreed that it should be a guiding principle for radioactive waste management organisations, and thus had ethical significance for their planning processes and day to day operations.

7.10 Reflections on the Method

This brief example and the map of judgements and principles shown in figure 7.1, give a flavour as to the use of the method in practice in structuring the deliberation of ethical issues in line with a reflective equilibrium approach. The content of the application of principles to judgements in this sketched out case, is perhaps to the eyes of a trained philosopher, rather simplistic. However, there are pragmatic benefits to the exercise, both for decision-making and for the participants themselves.

Partly given resource constraints, these workshops took place on a single day without prior participant involvement, training or expert-level information provision as might be expected in Government-run community and stakeholder engagement processes such as GM Nation?. This poses challenges in terms participant knowledge (from pre-workshop personal research on the issues, and the involvement of expert testimony in framing the terms of the deliberative engagement). It also raises issues of competency for novice deliberators to 'do' practical philosophy. This was one of the primary experimental outcomes of these workshops, to test whether lay participants could deliberate on ethics in a satisfactory and philosophically sound manner, without input from expert ethicists. Dreyfus and Dreyfus (2004) discuss the competency of individuals to undergo the evaluation of ethical issues. Under their terminology, the participants were novices, in the sense that they lacked prior experience of the necessary maxims or rules (i.e. knowledge of ethical principles, or the terminology of normative ethical judgements) needed to evaluate the issues that they were deliberating upon without help and active facilitation. As ethics terminology was unfamiliar to most participants, their wielding of these principle concepts was perhaps not complex or philosophically sophisticated, though it did involve reasoned opinion expression, was relevant to the source materials and topic focus, and involved exchange of ideas around ethical issues in a logical and structured manner. Although it cannot be expected that an individual's ethical competency should improve dramatically throughout a single day workshop, it was recognised that the quality of deliberation is partly related to issues of comfort and participant satisfaction in the process (Halvorsen 2001); and relaxed social interaction amongst participants. If participants are comfortable discussing issues amongst their peers and the facilitator, with their personal needs catered for, a sense of joint ownership in the process and fair consideration of their perspectives as equals, then there is evidence to suggest that this improves their level of confidence with expressing themselves in deliberative forums; engaging in technical (and thus also ethical) issues and hence improving the quality of the dialogue (Lindskold 1983).

The fact that topics of ethical interest identified by the participants closely matches that identified in the academic (Hadjilambrinos 1999, 1990; Shrader-

Frechette 1991; Shrader-Frechette 1999; Cotton 2009; Brook 1997) and grey lite-ratures (Damveld 1992; Timmerman 2003; Blowers 2006; Rawles 2000; Rawles 2004) on RWM ethics such as compensation/community benefits, the fairness of health risk burdens and decision-making involvement, is some testament to the competency and 'deliberative capacity' of diverse groups to engage with relevant ethical issues in a logically structured manner, maintained by the 'talk-centric' (Bohman and Rehg 1997) combination of participant-led judgements and principle selection. There is also evidence that participants displayed commitment and en-gagement in reflecting upon the relative value of one another's ideas, and this is an issue central to the success of ethical deliberation. In the workshops there was sufficient evidence of exchange of ideas rather than simply the 'top-of-the-mind' offhand views characteristic of shorter focus groups. The outcomes of the work-shops are not just bottom-up policy objectives, but also better informed judge-ments illustrated by a qualitative transformation in the direction of the dialogue. With successful facilitation, the reflective ethical mapping process allows oppor-tunities for collaborative learning, rather than encouraging one viewpoint to over-ride another. In this regard, the moral decisions that emerge as a context of these discussions take into account both theory driven principles as well as emotions, values and personal beliefs. Thus they can be considered truly deliberative, as Gracia (2003) attests - deliberation is the process in which those concerned by the decision are considered valid moral agents, obliged to give reasons for their own points of view and to listen to the reasons of others:

> "…in many cases the members of a group deliberation will differ in the final solution of the case, but the confrontation of their reasons will modi-fy the perception of the problem of everyone… Our moral decisions can-not be completely rational, due to the fact that they are influenced by feelings, values, beliefs, etc., but they must be reasonable, that is, wise and prudent. Deliberation is the main procedure to reach this goal. It ob-liges us to take others into account, respecting their different beliefs and values and prompting them to give reasons for their own points of view."

Though in some respects the judgements remain philosophically simple and straightforward, the goal is not just to display expert competency in ethical analy-sis, and hence build a consensus on what should be done. Rather it fulfils the goal of enriching the individual participants' own point of view with that of the others, increasing the maturity of the decision in and making it more wise or prudent. Though the deliberation on ethics has value for encouraging collaborative learning on the ethical issues, the final phase of the workshop provides greater clarity for third party evaluation of the decision options presented in light of the deliberative analysis of ethical issues.

7.11 Turning Issues into Courses of Action

The final phase of the reflective ethical mapping process involves closing down the deliberation to bring the philosophical reflection and group discussion back into the participatory-deliberative decision-making process and ensure the strong social democratic control of technology. Issue identification and action planning is the aim of the final phase.

7.11.1 Issue Identification

The identification of issues is proposed first as a listing or brainstorming exercise to draw out potential strategies for policy and practice that are contextually relevant to the foregoing discussion of ethically informed courses of action. In practice, participants are instructed to re-read the outputs of the reflective equilibrium stage of the workshop (coherently balanced judgements in light of principled perspectives and situated principles in light of case specific judgements). They are then charged with discussing how the ethical reflections drawn from the discussions might be borne out in a real decision, creatively imagining potential solutions to the ethical problems that have been identified and the steps that can be taken to ensure that technology decisions are ethically robust, thinking specifically about the involvement of different stakeholder groups.

The process begins with the identification of problems, which are listed along the left hand side of an action planning table. These problems can be summaries of the ethical issues identified in previous rounds, or else move beyond the previous discussions to present new ideas or problems. When listing the ideas, however, each participant must state why it is chosen, with reference to the preceding discussions. This helps to maintain a coherent link between prior rounds of discussion and the final options for consideration.

Following the listing exercise, a second brainstorming idea generation activity is designed to stimulate discussion of options and strategies that provide potential solutions to the problems identified. Again, these must have an ethical quality to them, when individuals state what the solutions could be, facilitators can then encourage the group to reflect upon the ethical character of these options in relation to the previous discussions. Participants are instructed to identify a list of those that might be responsible and those that might be affected by these problems and the potential solutions, followed by two further columns giving positive and negative justifications for putting the action into practice. This simplified model of identifying 'goods' and 'bads' allows encourages participants to reflect upon the implicit ethical foundations of policy strategy and political agency – requiring them to imagine the futures that they create through their actions and plans, and how these might affect different individuals within society. This has its roots in the Deweyan concept of teleological moral empathy, in the sense that it involves imaginative deliberation on the outcomes of particular courses of action. It also bears similarity with Rawls's Original Position, in that participants must consider the outcomes of strategies not in terms of personal gain, but rather from the perspective of others in civil society that might be affected. Once this is complete, the proposed actions are given a label and assigned a letter or number to delineate each option, followed by a 'weight' column which is initially left blank. Thus the table columns follow this format:

- Problems
- Options/strategies
- Who is responsible and who is affected
- Positive aspects
- Negative aspects
- Label
- Weight

The brainstorming and listing exercise is valuable in that it lays bare the outcomes of the decision-making process, opening up the ethical deliberation to reflection on practical matters. This has pragmatic value, as it lays the groundwork for action planning, and is likely to be persuasive to policy-makers due to the grounded policy-facing nature of the ethical assessment (see for example Light 2003).

7.11.2 Weighting Target

A further closing down mechanism is then needed in order to choose between the different options and identify the solutions that are desired by participants on the basis of the evaluation of ethical content and context throughout the workshop. This requires weighting and deciding mechanisms to reduce the number of potential options for further examination and implementation in policy. The basis for including a final set of methods to close down the workshop is conceptually grounded in multi-objective decision support (MODS) analysis. Simply put, MODS facilitates identification of an option or alternative from those that meet a range of different objectives (Hajkowicz and Prato 1998), rather than assessing the criteria that meet a single objective (Nijkamp 1989). With the focus upon multiple objectives, MODS are compatible with exploratory, bottom-up ethical deliberation intended in these workshops because they identify and then realise the means to achieve a range of options, rather than appraising different criteria for preselected options (which would produce bias through a framing effect on the decision-making process). From Howard (1991), and Hajkowicz and Prato (1998) it is possible to identify a generic model of MODS as:

- Defining the objectives
- Choosing the attributes
- Specifying the alternatives
- Transforming the attribute scales into commensurable units
- Assigning weights to the attributes which reflect their relative value to the decision maker
- Selecting and applying an algorithm for ranking the alternatives
- Choosing an alternative

The tool presented in this final phase is a simple weighting and scoring model that shares some similarities with the above approach:

- Discussion, suggestion and recording of ethically informed objectives or alternatives
- Identification of responsible or affected stakeholder groups/decision bodies
- Identification of positive and negative implications of implementing objectives
- Assign category labels
- Score objectives
- Calculate scores
- Reflect upon highest (and lowest) scoring objectives

The intention in implementing this method is to get participants to identify a range of ethically informed strategies, options or objectives that they consider worthy of further investigation in future workshops or other deliberative engagement forums. By asking them to consider the discussions they have had over the day, they are then asked to put forward what they feel were viable means to achieve the ethical goals identified throughout the session. The use of numbered weights to then prioritise amongst these different strategies, options or objectives has its roots in a number of other multi-criteria type approaches. Although comparatively simple when compared to other scoring approaches used in the various methods for choosing amongst radioactive waste management options (for example Atherton and French 1998; Chilvers et al. 2003; Burgess et al. 2004; Greenberg et al. 2002), a simplified visual weighting system first identifying the perceived benefits and drawbacks of each strategy and then scoring it, provides a simple tactile and visual approach that can easily be implemented with novice practitioners.

One means to achieve this is to use a target approach. The ideas/options are assigned letters and these are copied onto a 5 ringed target, which is intersected into the number of slices equivalent to the number of letters (options/ideas). Participants are again given a number of sticky dots equal to the number of letters and then asked to vote on each issue from 1-5 and place the dot closer to the centre to represent an idea meriting further exploration by decision-makers and dots on the edge for those deemed less important or impractical. There was no specific rank ordering process, participants were free to put them all on '5' or all on '1', but each was only allowed one vote per idea/option. The targeting method provides a clear, visual alternative to nominal group technique or Likert-scale type questionnaires or other similar voting or ranking procedures. It allows any number of options to be considered and provides the means for transparent dissemination by the participants. The target scores are used as a means for defining weights to the different options/ideas, whereby those with the highest weighting are taken forward to the final closing down session of the workshop, giving participants an opportunity to reflect upon the group's chosen ideas/options and feedback about their experience of the workshop, the development of the discussions (and their own understanding of the issues) after completion of the process.

7.11.3   Practical Summary of the Issue Identification and Weighting Target

- Total time taken for method 1 hr 20 minutes.
- Draw up the table and introduce the aims (10 minutes).
- Identify through group discussion and brainstorm a long list of potential options/strategies (20 minutes).
- Discuss in groups the stakeholders involved, and evaluate the positive and negative implications (20 minutes).
- Introduce the weighting target and hand out sticky dot 'votes' (10 minutes).
- Use nominal group technique voting on weighted target (15 minutes).
- Add up scores and identify 'winners' (5 minutes).

7.12   Example of the Method in Practice

Table 7.2 shows an example of the listing table that draws together the identified strategies and the evaluation of their feasibility in light of practical and ethical criteria. Table 7.3 shows the weighting of each of the options. In this workshop there was a clear consensus that emerged, as options H, I and J were equally scored with maximum weighting, implying that these three were the issues deemed most important for future options scoping and deliberation. What is interesting to note is that the three highlighted strategies all concerned political decision-making processes for radioactive waste, namely the power of community veto (termed a right of withdrawal), a concern for community rights and the examination of the impacts of a non-consultative decision on local communities, and thirdly concerns over political stability over long time frames, and finding ways to ensure the rights of future generations. Together these issues represent a concern with procedural fairness in decision-making, an issue which has been shown to heavily influence community perceptions of the acceptability of project siting outcomes (Gross 2007). Their appearance here reinforces the need for radioactive waste management organisations to ensure fair and transparent involvement of project site communities in decision-making processes over siting, not just now, but over multi-generational time frames (Fuji Johnson 2006).

Table 7.2 Options and weighting scheme

	Option/strategy	Who is responsible? Whom does it affect?	Evaluation		Weight (from target)
			Positive aspects	Negative aspects	
A	Reopen the disposal options debate, specifically focusing upon disposal of encased wastes in shallow, under sea mine shafts off the Hartlepool coast	Safeguarded by water, and so in nobody's 'back yard'.	No individual is affected	No means of checking if the waste has leaked It's potentially everyone's problem	17
B	Examining 'best practice' among radioactive waste management organisations internationally and following their example	International collaboration between scientific agencies	Get the best available advice Waste can be exported to countries where they can manage it more safely	Exporting the problem UK waste, UK should deal with it	23
C	Base siting decisions entirely upon the input of impartial scientific experts	Scientists and other experts drive the decision – discarding the local people's views Decision made on the best scientific evidence	Avoids community competition Money thrown at the problem will solve/alleviate it	It's tax payer's money – what else could we spend it on?	23
D	Destroy the waste – advance research into partitioning and transmutation and other possible waste reduction measures	Physicists Future generations	No more problem	Costly, can it ever be achieved?	19
E	Engage in protest actions due to mistrust of the local decision-making authorities (specifically the local council)	Local council is not trusted (spineless, incompetent)	Community voices heard	We will be ignored	16

Table 7.2 (*continued*)

F	Engage in a local referendum exploring and voting upon potential waste management strategies among community members	Local people Politicians	Understand the views of the community, not what politicians think they are	Goes around political representatives, so is it legitimate? Can the government override a 'no' decision?	24
G	Volunteerism – define the boundaries around which 'a community' is defined and establish who can be included in a volunteer decision	"Teesside city region" Which group should make the decision? The North East Regional Assembly	Needs a substantial cash incentive/benefits package	Size of the problem may change Location and geographical regions might not be the best/safest for waste disposal	15
H	Veto powers - establish the stage at which a right to withdraw from siting decisions is possible	Local people	Not forced to accept something we don't want	At what point does it 'click in'/become available?	35
I	Examine the impacts of top-down central government decision without consultation versus local decision-making control and the strengthening of legal protection/community rights	Politicians promising to uphold the consultation process outcomes	Views of the people are listened to	Could be overridden in the future with a change of government	35
J	Examine the feasibility of attempting to create long-term political structural stability in UK society and the host community – also educate future generations about RWM options and the ethical responsibilities of long-term waste stewardship	Future generations	Educating future generations to find the most adaptable solution to their needs	Impossible to do! Cannot be guaranteed.	35

Table 7.3 Scoring and weighting of options

Idea	1	2	3	4	5	Total score
A	4	0	1	0	2	17
B	1	2	0	2	2	23
C	2	1	0	1	3	23
D	1	3	1	1	1	19
E	4	1	0	0	2	16
F	1	0	2	3	1	20
G	5	0	0	0	2	15
H	0	0	0	0	7	35
I	0	0	0	0	7	35
J	0	0	0	0	7	35

7.12.1   Reflections on Issue Identification and Weighting Target

The value of this method lies in the ability to close down the discursive element of the workshop, and to once again ground the discussion of ethics in the context of real world decision-making. By thinking back on the day's discussions, re-examining the output sheets and further facilitated discussion, participants are able to think creatively around the ethical challenges presented throughout and suggest ideas that could remedy problems or implement ethically informed objectives. By then scoring these items this provides a clear indication that their input was valuable, whilst providing fair and balanced outputs. Crucially this method doesn't involve rank ordering; the weighted scores for each option can be as high or low for each option as the participants feel is appropriate (between 1 and 5). Thus, if participants feel that all objectives are equally important (or unimportant) for further investigation, they can use the votes accordingly. The weighted scores are intended to be discussed and reflected upon; they present a snapshot of the group's valuation of each of the ideas presented, rather than a formal mathematical model for deciding between options. The scores are therefore intended to be illustrative for further group discussion and reflection, rather than factor weights for an

MCDA type approach. This is coherent with the primary objective of the REM approach to improve the quality of ethical deliberation, rather than simply trying to select an option from a predefined set, or to enforce a consensus when none emerges.

7.13 Conclusions

Together these methods describe a process of ethical evaluation that is both deliberative and evaluative in scope. By using image based methods to structure imaginative scenarios and problem formations, charrette techniques to elicit judgements and the conceptual mapping of judgements to principles and vice versa, it is through this process that reflective equilibrium is achieved.

References

Alexander, T.: On Dewey and the Moral Imagination: Beyond Putnam and Rorty Toward a Postmodern Ethics. Transactions of the Charles S. Peirce Society 29(3), 369–400 (1993)

Atherton, E., French, S.: Valuing the Future: A MADA Example Involving Nuclear Waste Storage. Journal of Multi-Criteria Decision Analysis 7, 304–321 (1998)

Beauchamp, T.L., Childress, J.F.: Principles of Biomedical Ethics, 5th edn. Aufl. Oxford University Press, New York (2001)

Blowers, A.: Ethics and Decision Making for Radioactive Waste. Committee on Radioactive Waste Management, London (2006)

Bohman, J., Rehg, W.: Introduction. In: Bohman, J., Rehg, W. (eds.) Deliberative Democracy: Essays on Reason and Politics. MIT Press, Cambridge (1997)

Brook, A.: Ethics of Wastes: The Case of the Nuclear Fuel Cycle. In: Cragg, A.W., Koggel, C.M. (eds.) Contemporary Moral Issues. McGraw Hill Ryerson, Toronto (1997)

Burgess, J., Chilvers, J., Clark, J., Day, R., Hunt, J., King, S., Simmons, P., Stirling, A.: Citizens' and Specialists' Deliberate Options for Mapping the UK's Legacy Intermediate and High Level Radioactive Waste: A Report of the Deliberative Mapping Trial. CoRWM PSE Working Group (June-July 2004)

Caspary, W.R.: Dewey on Democracy. Cornell University Press, Ithaca (2000)

Chilvers, J., Burgess, J., Murlis, J.: Securing Public Confidence in Radioactive Waste Management: Developing a Vision for a Process and Stakeholder Engagement. University College London, London (2003)

Clark-Ibáñez, M.: Framing the social world with photo-elicitation interviews. American Behavioural Scientist 47(12), 1507–1527 (2004)

Collinson, D.: Ethics and aesthetics are one. British Journal of Aesthetics 25(3), 266–272 (1985)

Cotton, M.: Evaluating the 'ethical matrix' as a radioactive waste management deliberative decision-support tool. Environmental Values 18(2), 153–176 (2009)

Damveld, H.: Nuclear Waste and Ethics (1992), http://www.mfgroningen.nl/Nuclearwaste.htm (accessed November 2011)

Dewey, J.: Logic: The Theory of Inquiry. Rinehart and Winston, New York (1938)

Dreyfus, H.L., Dreyfus, S.E.: The Ethical Implications of the Five-Stage Skill-Acquisition Model. Bulletin of Science, Technology & Society 24(3), 251–264 (2004)

Epstein, I., Stevens, B., McKeever, P., Baruchel, S.: Photo Elicitation Interview (PEI): Using Photos to Elicit Children's Perspectives. International Journal of Qualitative Methods 5(3), 1–9 (2006)

Fesmire, S.: John Dewey and Moral Imagination: Pragmatism in Ethics. Indiana University Press, Bloomington (2003)

Fuji Johnson, G.: Deliberative democracy and precautionary public reasoning: exploratory thoughts. Les Ateliers de l'ethique: la Revue du CRÉUM 1(1), 81–87 (2006)

Goodin, R.E.: Utilitarianism as a Public Philosophy. Cambridge University Press, Cambridge (1995)

Gracia, D.: Ethical case deliberation and decision making. Medicine, Health Care and Philosophy 6(3), 227–233 (2003)

Grassian, V.: Moral Reasoning: Ethical Theory and Some Contemporary Moral Problems. Prentice-Hall, Wilmington California (1981)

Greenberg, M., Burger, J., Powers, C., Leschine, T., Lowrie, K., Friedlander, B., Faustman, E., Griffith, W., Kosson, D.: Choosing Remediation and Waste Management Options at Hazardous and Radioactive Waste Sites. Remediation, 39–58 (2002)

Gross, C.: Community perspectives of wind energy in Australia: The application of a justice and community fairness framework to increase social acceptance. Energy Policy 35(5), 2727–2736 (2007)

Hadjilambrinos, C.: Nuclear Power as an Ethical Issue: Utilitarian Ethics and Egalitarian Responses. Bulletin of Science, Technology and Society 10, 286–288 (1990)

Hadjilambrinos, C.: Toward a Rational Policy for the Management of High-Level Radioactive Waste: Integrating Science and Ethics. Bulletin of Science, Technology & Society 19(3), 179–189 (1999)

Haidt, J.: The emotional dog and its rational tail: a social intuitionist approach to moral judgment. Psychological Review 108, 814–834 (2001)

Haidt, J.: The moral emotions. In Handbook of affective sciences. In: Davidson, R.J., Scherer, K.R., Goldsmith, H.H. (eds.) Handbook of Affective Sciences, pp. 852–870. Oxford University Press, Oxford (2003)

Hajkowicz, S., Prato, T.: Multiple Objective Decision Analysis of Farming Systems in Goodwater Creek Watershed. University of Missouri-Columbia, Missouri, Columbia (1998)

Halvorsen, K.E.: Assessing Public Partcipation Techniques for Comfort, Convenience, Satisfaction, and Deliberation. Environmental Management 28(2), 179–186 (2001)

Harper, D.: Talking about pictures: a case for photo elicitation. Visual Studies 17(1), 13–26 (2002)

Heinonen, T., Cheung, M.: Views from the Village: Photonovella with Women in Rural China. INternational Journal of Qualitative Methods 6(4), 35–52 (2007)

Higgins, J.M.: 101 Creative problem solving techniques. New Management Publishing, Florida (1994)

Howard, A.F.: A critical look at multiple criteria decision making techniques with reference to forestry applications. Canadian Journal of Forest Research 21, 1649–1659 (1991)

Hurworth, R.: Photo-Interviewing for research. Social Research Update: Sociology at the University of Surrey (40) (2003)

Kaler, J.: What's the Good of Ethical Theory? Business Ethics: A European Review 8(4), 206–213 (1999)

Light, A.: Moral and Political Reasoning in Environmental Practice. MIT Press, Boston (2003)

Lindskold, S.: Conflict and Conciliation with Groups and Individuals. In: Blumberg, H.H., Hare, A.P., Kent, V., Davies, M.F. (eds.) Small Groups and Social Interaction. John Wiley & Sons, Chichester (1983)

Makansi, J.: Lights Out: The Electricity Crisis, the Global Economiy and What it Means to You. John Wiley and Sons, London (2007)

McFadzean, E.: Improving Group Productivity with Group Support Systems and Creative Problem Solving Techniques. Creativity and Innovation Management 6(4), 218–225 (1997)

McVea, J.F.: Constructing Good Decisions in Ethically Charged Situations: The Role of Dramatic Rehearsal. Journal of Business Ethics 70, 375–390 (2007)

Michalko, M.: ThinkerToys: A Handbook of Creative Thinking Techniques. Ten Speed Press, Berkeley (2006)

Moberg, D., Seabright, M.: The Development of Moral Imagination. Business Ethics Quarterly 10, 845–884 (2000)

Nijkamp, P.: Multicriteria analysis: a decision support system for sustainable environmental management. In: Archibugi, F., Nijkamp, P. (eds.) Economy and Ecology: Towards Sustainable Development. Kluwer, London (1989)

Patterson, W.: Keeping the Lights On: Towards Sustainable Electricity. Earthscan, London (2007)

Proctor, T.: New Developments in Computer Assisted Creative Problem Solving. Creativity and Innovation Management 6(2), 94–98 (1997)

Prosser, J.: Image-Based Research: A Sourcebook for Qualitative Researchers. Routledge Falmer (1998)

Rachels, J.: The Elements of Moral Philosophy, 2nd Aufl. McGraw-Hill, New York (1993)

Rawles, K.: Ethical issues in the disposal of radioactive waste. United Kingdom Nirex Limited, Harwell (2000)

Rawles, K.: Ethical issues in the siting of radioactive waste management facilities: the role of local communities. United Kingdom Nirex Limited, Harwell (2004)

Satterfield, T.: In Search of Value Literacy: Suggestions for the Elicitation of Environmental Values. Environmental Values 10(3), 331–359 (2001)

Schmidt-Felzmann, H.: Pragmatic Principles - Methodological Pragmatism in the Principle-Based Approach to Bioethics. Journal of Medicine and Philosophy 28(5-6), 581–596 (2003)

Shrader-Frechette, K.: Ethical Dilemmas and Radioactive Waste: A Survey of the Issues. Environmental Ethics 13(4), 327–343 (1991)

Shrader-Frechette, K.S.: Expert Decision-Making in Risk Analysis: the Case of Yucca Mountain. In: Anderson, K. (ed.) Values in Decisions on Risk (VALDOR): A Symposium in the RISCOM Programme Addressing Transparency in Risk Assessment and Decision Making. Congrex, Stockholm (1999)

Smith, P.D., Emmison, M.J.: Researching the Visual: Images, Objects, Contexts and Interactions in Social and Cultural Inquiry. Sage (2000)

Timmerman, P.: Ethics of High Level Nuclear Waste Disposal in Canada. NWMO Background Papers (2003)

Vidal, R.V.V.: Creativity and problem solving. Economic Analysis Working Papers 3(14) (2004)

Werhane, P.H.: Moral Imagination and Management Decision-Making. Oxford University Press, Oxford (1999)

Werhane, P.H.: Moral Imagination and Systems Thinking. Journal of Business Ethics 38, 33–42 (2002)

Chapter 8
Conclusions

8.1 Introduction

In this concluding chapter I aim to review the development of this ethical decision-support procedure termed *Reflective Ethical Mapping* (REM) both in relation to the pragmatist conceptual framework that I outlined in chapter 3, and importantly, to the practice of participatory technology assessment (PTA).

The thematic issues raised within this book highlight the breadth and complexity of ethical considerations that lay citizens bring into discussions on the governance of socially and ethically contentious technologies (SECT), discussed through reference to the practical case study of managing long-lived radioactive wastes. The key meta-ethical position worth reiterating here is that cultural discourse on technological risks should not be constrained solely within scientific and technical analysis of health and environmental impacts, costs or safety. Debate about the far reaching consequences of technological development and implementation cannot be a purely objective and factual discussion, bounded by the rationality of techno-scientific analysis. Neither quantitative risk assessments alone, nor finding ways to encourage better public understanding of scientific and technical issues will facilitate consensus building or public acceptance SECT in the public realm, because the nature of risk debates implicitly involves complex ethical issues, numerous and conflicting relationships, trust and social capital. Public reactions to controversial technologies are driven by conflicting perspectives on governmental, industry and stakeholder obligations towards communities, environments and future generations, and our understanding of ethics must inevitably lead to negotiation between competing interests based on divergent ethical perspectives.

The cultural, linguistic and participatory-deliberative turns in technology policy have served to broaden the realm of technology management debates out from the narrow confines of techno-scientific analysis, quantitative risk assessment and the forecasting of technological trends. These factors have allowed the normative and deliberative competency of citizen perspectives to be taken seriously in key decision-making contexts. In this book I have sought to provide practical tools to support ethical value-based discussions by facilitating a deliberative process that gives relevant ethical arguments fair and balanced consideration. Thus, the

M. Cotton, *Ethics and Technology Assessment: A Participatory Approach*,
Studies in Applied Philosophy, Epistemology and Rational Ethics 13,
DOI: 10.1007/978-3-642-45088-4_8, © Springer-Verlag Berlin Heidelberg 2014

intention has been to contribute a novel methodological process to this fledgling field of ethical evaluation in participatory technology assessment (PTA) that not only adds to the theory of applied ethics in operationalising John Rawls's concept of *reflective equilibrium*, but also the practice of technology governance. My hope is that the application of the proposed reflective ethical mapping approach can facilitate better quality decisions over the management and implementation of SECT in society through explicit deliberation and reflection on ethical issues; and that it could be applied to other controversial policies where explicit ethical analysis is necessary to ensure the adequate social control of controversial technological programmes.

8.2 The Problem Focus

The research that underpins this book has taken the form of an experiment in practical philosophy, concerned with the application of both theoretical and practical thought with a view to action (Haldane 2012). In short, it is premised on the notion that not only should lay citizens somehow provide 'input' in the form of normative ethical values, but that they can, with no formal training in ethics, competently perform ethical evaluation in the form of an analytic-deliberative decision-support process which has practical policy implications. The supposition that grounded this methodological framework being that with the right tools and facilitator guidance, a lay citizen panel can perform the roles traditionally occupied by an ethics committee, forum or expert panel; moving through sequential stages of discussion to focus their analysis in a way that rivals formalised training in normative or applied ethics. The Reflective Ethical Mapping (REM) approach outlined in these chapters is therefore presented as a means to assist those who want to improve the quality of ethical deliberation by capturing a broad range of ethically relevant aspects of an issue, grounding them in a practical context and evaluating them both in light of the judgements they make and the principles that coherently 'fit'.

8.3 Ethical Tools

From early in the process of developing the reflective ethical mapping approach I envisaged the concept as an ethical toolbox or toolkit of methods suitable for PTA practice. To some extent this work sits alongside similar toolbox approaches that have gained popularity in the literature on ethics education and practical ethics, particularly those of Weston (2000) and Baggini and Fosl (2007). In both these examples, the concept of an ethical toolbox is a collection of thought procedures to tease out and evaluate ethical issues for different practical applications, with the emphasis upon individual learning, philosophical reflection and ethically informed decision-making. In contrast the Ethical Bio-Technology Assessment tools development process (Ethical Bio-TA project) (Kaiser et al.

2004) (of which some tools were assessed in chapter 4), sought to identify and test a series of existing practical and participatory methods in light of their potential contribution to ethical evaluation in different group-based policy making contexts, primarily around issues of governance in agriculture and food production (Forsberg 2007; Beekman and Brom 2007; Kaiser et al. 2004; Kaiser et al. 2007; Kaiser and Forsberg 2001). The substantive practical contribution that this book provides is something of a middle ground between these two approaches. It has specifically focussed upon ethical tool development, whereby tools are "…judgement aids that help justify value choices without recourse to substantive theories or value systems of limited scope" (Forsberg 2007). This has involved empirical testing of new group-based deliberative methods (in relation to the problem of radioactive waste management) but has also involved the development of a rationale or conceptual framework in which new techniques can be developed in the future.

Upon reflection on the practice of these deliberative workshops, it is clear that the toolkit metaphor has become somewhat redundant in the development of this approach. A toolkit implies distinct tools for different tasks, operating independently of one another. The REM approach has developed into a deliberative *procedure* operating sequentially, building ethical analysis through group discussion in discreet stages. Although it would be possible to substitute different tools and methods into the framework as the need arose, it is the sequential stages of issue mapping, judgement 'elicitation', principle based evaluation and creative problem solving that remain the essence of the REM approach.

8.4 The Ontology of Reflective Ethical Mapping

This approach is grounded within ontological (and hence meta-ethical) anti-foundationalism. A foundationalist ontology would tend towards a monistic meta-ethical claim that the proposed REM approach is the only appropriate set of methods to be used (or perhaps the best set of methods). Foundationalism and ethical absolutism have been rejected on the grounds that the preoccupation with general and abstract truths is counterproductive, in the sense that it distracts attention from concrete problems and conflicts tied to particular times, places and actors. I argue that ethical decision-making requires flexibility and context sensitivity to be successful. It is for this reason that a pragmatist approach has been adopted. Rather than arguing that these tools are the *right* ones to use, the intention has been to simply assess whether REM framework proves *useful*. The rationale for the development of ethical tools has thus been based upon what could be considered a non-reductionist ontology, in that it seeks to take into account a broad array of phenomena (including technical, scientific, principle and judgement-based factors) without reducing them to one or two core notions. I therefore attempt to present a methodology that attends to a great variety of human experiences rather than trying to find an underlying common phenomenon to explain 'how it all works'.

Where the pragmatism that informs this REM model differs from similar approaches (such as moral relativism) is in the *constructive* nature of the conceptual framework. The purpose of pragmatism's critique of traditional moral philosophy, in particular normative ethics, has been to open the way to new insight (Parker 1996) not merely deny that there can be any satisfactorily absolute moral answer. I have presented an argument that ethical action can only be discovered empirically through trial and error and that the morality of any evaluations that result are specific only to the particular situation, within a particular space and time. Thus, the goal has been to contribute the means to ethically evaluate SECT through exploring a complex network of techno-scientific and practical information and normative judgements and principles to generate practical solutions to real moral problems identified in a bottom-up manner.

The potential solutions derived from the workshops are not generalisable to all situations, they are specific, particular and open to reinterpretation and change. Similarly neither judgements nor principles are considered fixed, abstract or immutable; they are used as tools to evaluate the problems, not as ends in themselves (Farber 1999). The pragmatism I espouse also implies a commitment to reflective research - continually testing and shaping practice and theory, adding new tools and techniques or adapting existing ones. The programme of research therefore echoes Weston (2002), Mepham (2005) and Seedhouse (1998; 1988) when they assert that a tool is simply a means to assist ethical evaluation by clarifying thinking, argumentation and (in this case) embedding it within a specific policy-making context; rather than providing an assessment metric, substitute for critical thinking, or a new normative theory.

8.5   The Pragmatist Rationale for REM

As a project in philosophical pragmatism, practice become the primary mode of analysis, meaning that ethical evaluation can be found by paying attention to the practical consequences of theory (Rosenthal 1994), rather than defining the correct normative rules or principles through abstracted thought experimentation. Drawing upon the framework outlined by Keulartz et al. (2004) the role of traditional normative ethical theories has been repeatedly challenged throughout as being a limited conception of the ethical issues that technologies create. As shown in chapter 3, modern technological culture is dynamic in character. The development and implementation of new technologies alters social relationships within communities and so the ethical norms and values of a society are continually replaced as we are regularly confronted with novel moral problems. I have adopted this line of philosophical analysis; supporting the claim that traditional normative ethics is inherently 'technology blind' insofar as it places the rational moral actor in the centre and treats technologies as morally neutral tools in their hands.

Although pragmatist technology ethics may at first appear to be simply a blanket critique of normative theory-based approaches, in truth the challenge is more subtle.

New technologies cause new ethical problems to arise and the normative ethical frameworks that pre-date modern technological culture often simply lack the *vocabulary* to capture this dynamic character accurately (Keulartz et al. 2004). Technology challenges traditional ethical norms, impacting upon relationships among individuals and challenging how they deal with one another. For example, changes in the development and implementation of medical technology challenge traditional definitions of concepts such as *human life*; as illustrated in debates on contraception, abortion and euthanasia (Winston 2003); and similar conflicts have emerged over novel technologically defined moral problems such as those posed by genetically modified foods, xenotransplantation (organ transfer from animals to humans), or stem-cell research. As illustrated in the nuclear power and RWM examples used throughout this book, radiation risks similarly present novel moral challenges. As Weart (1988) argues, radiation is feared due to its propensity to transmute living organisms, and communities accepting nuclear facilities are often stigmatised because of this: perceived not simply as risky places, but as contaminating places – affecting perceptions of nuclear communities from within and without. Risk bearing technologies such as those related to the nuclear fuel cycle alter the social fabric of affected host communities and the relationships within and around them, which in turn has ethical ramifications that challenge the concept of technology as a morally neutral tool.

In relation to a pragmatist conception of technology ethics we must consider our moral theories, principles and personal judgements as methods of justification for evaluating ideas, seeking to understand the nature of ethical situations in order to see how they are constructed and contested; relying upon empirically given phenomena to search for useful generalisations and explanations. Iterative deliberation among affected citizens using REM is the proposed means to produce the aforementioned contextually relevant 'moral vocabulary' to accurately describe and assess the ethical issues inherent to the management of SECT; developing new terminologies of principles and judgements that are contextualised by practical matters (such as techno-scientific objectives and considerations), the normative competence of citizens in defining rational (and non-rational) moral judgements and their relationship to the plurality of citizen-stakeholder perspectives and values. Rather than applying pre-given normative rules or maxims to a practical situation, the more bottom-up REM approach allows the exploration of new possibilities by highlighting the practical and creative character of finding solutions to moral problems.

As previously stated, the REM approach is intended to operate as a multi-staged 'procedure'. This procedure has, purely for practical reasons, been condensed for practicality into a day-long process for structuring ethical deliberation. The one-day workshop format has the advantage of reducing costs and other resource constraints, however, in many Technology Assessment processes decision-making is likely to occur over much longer periods of time. It would likely prove useful to develop an iterative process of ethical deliberation that is spread across multiple meetings or venues; either by allowing time for reflection between the separate stages of the REM approach, or by repeating and refining the issue through multiple iterations in consecutive workshops. This would allow time for participants to consider their inputs at different stages, reflect upon one another's values and develop greater competencies at ethical evaluation using the principle based terminology inherent to reflective equilibrium.

8.6 Practical and Empirical Considerations

When looking at some of the outputs presented in chapters 6 and 7, some may wonder if the citizen participants had the opportunity or capacity to develop what could arguably be termed sufficient *ethical evaluative competence* to make full use of the methods. Some such as Giovanni (2012) have suggested that deliberative competence, the capacity of individuals to rationally evaluate the complexities of the technology in question, should be the criterion through which we evaluate the efficacy of deliberative methods. However, I adopt a rather more optimistic stance, suggesting that rather than reliance upon *expertise* as the basis of sound assessment, we should instead focus upon the involvement of lay citizen participants and their *normative competency*, to borrow Davies and Burgess's (2004) term, or perhaps more specifically an *ethical communicative competency*. By this I mean that the emphasis should be upon the capacity of both the process and the individuals involved to draw upon technical, social and ethical criteria in constructing their arguments about RWM strategy. Thus it is their capacity to articulate and communicate their values clearly, rather than their expert knowledge or ability to wield ethical concepts, that is the important facet of the deliberative process. This is something of a controversial point, as the issue of technical competency is often considered central to the success of deliberative methods and processes, whereby *competent* participants are judged to have particular rational capacities and abilities that legitimate decision-outcomes. These competencies include the rationality and capacity of participants to seek consensus on the procedures that they want to employ, articulate and criticise factual claims on the basis of the "state of the art" of scientific knowledge and other forms of problem-adequate knowledge, interpret factual evidence through analytical reasoning, disclose their relevant values and preferences, process data, arguments and evaluations in a structured format (Renn and Webler 1995; Jaeger et al. 2001; Renn 1998).

Translating these concepts across to the aforementioned problem of ethical evaluative competence; the workshops displayed numerous examples whereby participants showed the ability to assess aspects of the socio-technical issues in terms of their own moral perspectives, communicate this to one another and to listen, consider the moral perspectives of others, using the methods to both record statements and conceptually structure the deliberation. The aspect of the process that proved most challenging for participants was the use of principle-based terminology because they often lacked familiarity with the concepts and the essentially intellectual procedures involved in their application to ethical problems. Therefore, they often failed to distinguish specifically ethical issues from technical, social or philosophical questions or concerns in the way that experienced ethicists might be expected to do. In some respects the focus upon practical and social matters was a benefit, as it served to contextualise the problem with background, non-ethical concerns and thus come closer to realising Daniels's description of wide reflective equilibrium (Daniels 1979). One less desirable consequence, however, was that the different types of question were sometimes confused or conflated in the deliberative process, undermining the clarity of ethical discussion for other group members.

However, with sustained involvement using the tools over a longer period of time, citizen-stakeholders would gain competence in differentiating ethical from non-ethical matters and develop greater aptitude in relating one to the other. This problem was largely related to the practical constraints of working with 'novice', to use Dreyfus and Dreyfus's (2004) term, volunteers, without specialist support or sufficient prior information provision or exposure to the ethical tool based approach. As previously stated, the use of REM in the context of a long-term deliberative engagement process between the decision-makers and local community representatives would alleviate this problem. Such a process would entail a structured programme of constructive participant learning, and hence support the development over time of higher levels of ethical evaluative competency. In relation to this it is necessary to ground this proposed REM approach within the broader fields of analytic-deliberative methods and thus to reflect on the means to alleviate the problems of insufficient ethical evaluative competence.

8.7 Re-engaging Citizens with Specialists

Though the core argument is that ethicists and specialists on ethics panels and committees lack any special normative competency when compared to lay citizens, what they do possess, however, is greater experience in wielding theory to bear on practice and thus higher levels of ethical evaluative competency. Despite continued attempts throughout the REM development process to ease participants through the transition from thinking in concrete to more abstract terms, clearly there are some issues remaining which may be alleviated by incorporating ethical specialist support for citizens throughout the process. Guidance in this area comes from established analytic-deliberative methods such as Deliberative Mapping (DM), where scientific and technical specialists have been used to support citizen deliberation. In DM trials, specialists were first interviewed and their option assessment preferences recorded (using multi-criteria mapping software). In the subsequent workshops with citizens, specialists took part and supported them through the option assessment process (Burgess et al. 2004). The evidence from DM studies suggests that using ethical specialists in a similar support role could prove useful in future REM workshops. In the DM trial for RWM, the support role of specialists involved assisting in the initial provision of information to citizen participants, joining them in determining what each RWM option meant (including the technical elements and other social, economic, environmental and political implications), handling questions and comments informally during the meetings, providing additional information and participating in structured 'conversations' with pairs of citizens where each individual specialist was questioned about the options for managing radioactive waste (Burgess et al. 2004). Adopting a similar approach to specialist input may therefore alleviate the difficulties encountered by lay participants, giving them more guidance and information and helping them to distinguish technical from ethical aspects with greater ease. Nevertheless, this does not negate the normative contention of this book, that the elicitation of ethical judgements and selection of ethical principles should remain participant-controlled in order to satisfy the bottom-up legitimacy of the ethical evaluations argued for throughout.

8.8 Application of the REM Approach to Decision-Support

This work has primarily focussed upon the development of a deliberative ethical assessment framework and the testing of new methodological approaches. Questions remain however over the translation of these new methods from a pilot study to real world policy-making. The research was hypothetical in nature, whereby participants had to imagine that their community would be chosen for a waste facility, rather than a real decision process in which the participants had an equally real interest, limiting the practice-focus needed to fully evaluate the REM approach. Despite this, however, guidance for integrating REM as a PTA decision-support tool is needed. One simple means to utilise REM could simply be as a one-off event, used in concert with other deliberative methods such as citizens' juries or consensus conferences. In the radioactive waste management example it is likely that ethical issues will arise through the deliberative engagement process with communities integral to the proposed partnership approach espoused by the UK government. REM is flexible in that it can provide an opportunity for stakeholders and citizens to consider ethical issues involved at the different decision points. The method could be used either as a one-off workshop to raise and evaluate ethical issues or ideally over a longer period, allowing participants to undergo a learning process and improve their competence in ethical evaluation. By doing so, improving the ethical competence of lay participants would make practice more 'intelligent' (Lekan 2006, Winston 2003), i.e. sensitive to carefully evaluated ethical concerns and justifications thus strengthening the ethical validity of the decision-making process.

8.9 Conclusions

The reflective ethical mapping approach (REM) provides a methodological toolkit or decision-support procedure that is bottom-up, participant led and coherentist in its approach and thus has legitimacy irrespective of the capacities of the individual participants for philosophical reflection. REM provides the means to first generate discussion about practical, technical and political matters; identify a series of actants, relationships between disparate elements within an actor network 'map' in terms of cause and consequence; reflect upon individual judgements about the ethical issues raised, and reformulate these judgements in light of ethical principles;

recontextualise principles in light of specific judgements and cases, and then draw the deliberation to a close by imagining future courses of action and choosing between them based upon personal preference grounded in deliberative ethical competence and social learning about the ethical considerations throughout. The methodological development of the REM approach as a sequential procedure for the consideration of ethics is not fixed or closed, not exclusively the purview of the professional ethicist, and open to any and all that wish to take part. The epistemological value of the REM approach principally lies in its structure as a coherentist model of ethical assessment that has both opening up and closing down mechanisms suitable for analytic deliberative decision-making as part of a PTA process. These methods outlined here are, however, illustrative rather than prescriptive. Other tools can be added or subtracted based upon case considerations, pilot testing, practitioner judgement and participant feedback; making this is a truly pragmatic project - open ended, case specific and based upon empirical testing and experience.

References

Baggini, J., Fosl, P.S.: The Ethics Toolkit: A Compendium of Ethical Concepts and Methods. Wiley-Blackwell, London (2007)

Beekman, V., Brom, F.W.A.: Ethical Tools to Support Systematic Public Deliberations About the Ethical Aspects of Agricultural Biotechnologies. Journal of Agricultural and Environmental Ethics 20(1), 3–12 (2007)

Burgess, J., Chilvers, J., Clark, J., Day, R., Hunt, J., King, S., Simmons, P., Stirling, A.: Deliberate options for managing the UK's legacy intermediate and high level radio-active waste: a report of the Deliberative Mapping Trial. CoRWM PSE Working Group, London (June-July 2004)

Daniels, N.: Wide Reflective Equilibrium and Theory Acceptance in Ethics. Journal of Philosophy 76(5), 256–282 (1979)

Davies, G., Burgess, J.: Challenging the 'view from nowhere': citizen reflections on specialist expertise in a deliberative process. Health and Place 10(4), 349–361 (2004)

Dreyfus, H.L., Dreyfus, S.E.: The Ethical Implications of the Five-Stage Skill-Acquisition Model. Bulletin of Science, Technology & Society 24(3), 251–264 (2004)

Farber, D.A.: Eco-pragmatism: Making Sensible Environmental Decisions in an Uncertain World. University of Chicago Press, London (1999)

Forsberg, E.M.: Pluralism, the ethical matrix, and coming to conclusions. Journal of Agricultural and Environmental Ethics 20(4), 455–468 (2007)

Giovanni, B.: The Art of Deliberating: Democracy, Deliberation and the Life Sciences between History and Theory. Springer, Berlin (2012)

Haldane, J.: Practical Philosophy: Ethics, Society and Culture. Imprint Academic, Exeter (2012)

Jaeger, C., Renn, O., Rosa, E., Webler, T.: Risk, Uncertainty and Rational Action. Earthscan, London (2001)

Kaiser, M., Forsberg, E.M.: Assessing Fisheries - Using an Ethical Matrix in Participatory Processes. Journal of Agricultural and Environmental Ethics 14, 191–200 (2001)

Kaiser, M., Millar, K., Forsberg, E.-M., Baune, O., Mepham, B., Thorstensen, E., Tomkins, S.: Decision-Making Frameworks. In: Beekman, V. (ed.) Evaluation of Ethical Bio-technology Assessment Tools for Agriculture and Food Production: Interim Report Ethical Bio-TA Tools. Agricultural Economics Research Institute, The Hague (2004)

Kaiser, M., Millar, K., Forsberg, E.M., Thorstensen, E., Tomkins, S.: Developing the ethical matrix as a decision support framework: GM fish as a case study. Journal of Agricultural and Environmental Ethics 20(1), 53–63 (2007)

Keulartz, J., Shermer, M., Korthals, M., Swierstra, T.: Ethics in Technological Culture: A Programmatic Proposal for a Pragmatist Approach. Science, Technology & Human Values 29(1), 3–29 (2004)

Lekan, T.: Pragmatist Metaethics: Moral Theory as Deliberative Practice. Southern Journal of Philosophy 44(2), 253–272 (2006)

Mepham, B.: Bioethics: An introduction for the biosciences. Oxford University Press, Oxford (2005)

Parker, K.A.: Pragmatism and Environmental Thought. In: Light, A., Katz, E. (eds.) Environmental Pragmatism. Routledge, London (1996)

Renn, O.: The role of risk communication and public dialogue for improving risk management. Risk Decision and Policy 3(1), 5–30 (1998)

Renn, O., Webler, T.: Fairness and Competence in Citizen Participation, Technology, Risk and Society. Kluwer Academic Publishers, Dordrecht (1995)

Rosenthal, S.B.: Charles Pierce's Pragmatic Pluralism. State University of New York, New York (1994)

Seedhouse, D.: Ethics: the Heart of Health Care. Wiley, Chichester (1998)

Seedhouse, D.: Ethics: the Heart of Health Care. John Wiley and Sons Ltd., Chichester (1988)

Weart, S.R.: Nuclear Fear: A History of Images. Harvard University Press, Cambridge (1988)

Weston, A.: Toward A Really Practical Ethics. In: The Society for the Advancement of American Philosophy: Annual Conference, Portland, Maine (2002)

Weston, A.: A 21st Century Ethical Toolbox. Oxford University Press, Oxford (2000)

Winston, M.: Children of invention. In: Winston, M., Edelbach, R. (eds.) Society, Ethics, and Technology. Wadsworth, Belmont (2003)

Subject Index